Applications Manual

Differential Equations

AND BOUNDARY VALUE PROBLEMS

COMPUTING AND MODELING

FOURTH EDITION

Differential Equations

COMPUTING AND MODELING

FOURTH EDITION

C. Henry Edwards
David E. Penney
The University of Georgia

With the assistance of
David Calvis
Baldwin-Wallace College

Upper Saddle River, NJ 07458

Editorial Director, Computer Science, Engineering, and Advanced Mathematics: *Marcia J. Horton*
Senior Editor: *Holly Stark*
Editorial Assistant: *Jennifer Lonschein*
Senior Managing Editor: *Scott Disanno*
Production Editor: *Irwin Zucker*
Supplement Cover Designer: *Daniel Sandin*
Manufacturing Buyer: *Lisa McDowell*

Pearson Prentice Hall
Pearson Education, Inc.
Upper Saddle River, NJ 07458

Printed in the United States of America

10 9 8 7 6 5 4 3 2 1

ISBN 0-13-600679-5
978-0-13-600679-4

Pearson Education Ltd., *London*
Pearson Education Australia Pty. Ltd., *Sydney*
Pearson Education Singapore, Pte. Ltd.
Pearson Education North Asia Ltd., *Hong Kong*
Pearson Education Canada, Inc., *Toronto*
Pearson Educación de Mexico, S.A. de C.V.
Pearson Education—Japan, *Tokyo*
Pearson Education Malaysia, Pte. Ltd.
Pearson Education, Inc., *Upper Saddle River, New Jersey*

Contents

5 Linear Systems of Differential Equations

6 Nonlinear Systems and Phenomena

7 Laplace Transform Methods

8 Power Series Methods

Preface

This applications manual is provided to support the inclusion of computing experiences in the introductory differential equations course. It's four dozen projects are numbered in correspondence with the section-ending "application modules" in the textbook Edwards & Penney, **DIFFERENTIAL EQUATIONS AND BOUNDARY VALUE PROBLEMS: Computing and Modeling**, 4th edition, Prentice Hall (2008). The textbook applications are expanded here and augmented with detailed discussion and illustration of applicable *Maple*, *Mathematica*, and MATLAB techniques.

Each project typically begins with a technology-neutral discussion of the investigation at hand. This introductory general discussion is then followed by separate **Using *Maple***, **Using *Mathematica***, and **Using MATLAB** sections that can be read independently. Thus we offer three separate computing threads that can be followed through the manual, depending on the computing environment chosen for use. However, a student following one thread may well benefit from browsing the others to compare merits and styles of different computational systems. In my own teaching I have experimented with different possibilities, including

- All students in a class using the same computing system;
- Different students in the same class using different systems;
- Students using either *Maple* or *Mathematica* for symbolic computation and MATLAB for numeric computation.

Each project is designed to provide the basis for an outside-class assignment that will engage students for a period of several days (or perhaps a week or two). Each project is supported by a *Maple* V worksheet, a *Mathematica* notebook, and a MATLAB m-file that illustrates the computing techniques used in the project. Students can download these technology-specific versions of the individual projects from the DE projects page at the web site listed below. A downloadable text version of each section of this manual is also provided so that students can copy-paste commands directly from the text into an M/M/M window for immediate execution as they read this manual interactively.

This new edition of the manual includes ten "Further Investigations" that suggest more conceptual and open-ended student explorations. These new additions, one per chapter in the textbook, appear at the ends of Applications 1.4, 2.1, 3.1, 4.2, 5.2, 6.5B, 7.2, 8.2, 9.5, and 10.5A.

Henry Edwards
David Calvis

h.edwards@mindspring.com
dcalvis@bw.edu

www.prenhall.com/edwards

Chapter 1

First-Order Differential Equations

1.3 Application
Slope Fields and Solution Curves

A number of specialized differential equations packages are available as freeware or shareware. Links to download sites offering such software packages are provided on the Web page **www.prenhall.com/edwards** that supports this text and manual. Such systems automate the construction of direction fields and solution curves, as do some graphing calculators.

Computer algebra systems and technical computing environments such as *Maple*, *Mathematica* and MATLAB provide extensive resources for the study of differential equations. For instance, the *Maple* command

```
with(DEtools):
DEplot( diff(y(x),x)=sin(x-y(x)), y(x),
        x=-5..5, y=-5..5 );
```

and the *Mathematica* command

```
Needs["Graphics`PlotField`"]
PlotVectorField[{1, Sin[x-y]}, {x,-5,5}, {y,-5,5}]
```

produce slope (or direction) fields similar to the one shown in Fig. 1.3.25 of the text.

Figure 1.3.25 itself was generated by the MATLAB program **dfield** (John Polking and David Arnold, *Ordinary Differential Equations Using MATLAB* (2nd edition), Prentice Hall, 1999) that is available free for educational use. When a differential equation is entered in the **dfield** setup menu (as illustrated in Fig. 1.3.26 of the text), you can immediately plot a direction field and then — with a single mouse click — plot also the solution curve through any desired point.

For example, the figure at the top of page 2 shows a slope field and typical solution curves (generated using **dfield**) for the differential equation $y' = x - y$. It appears that there exists a (single) *straight line* solution curve that all other solution

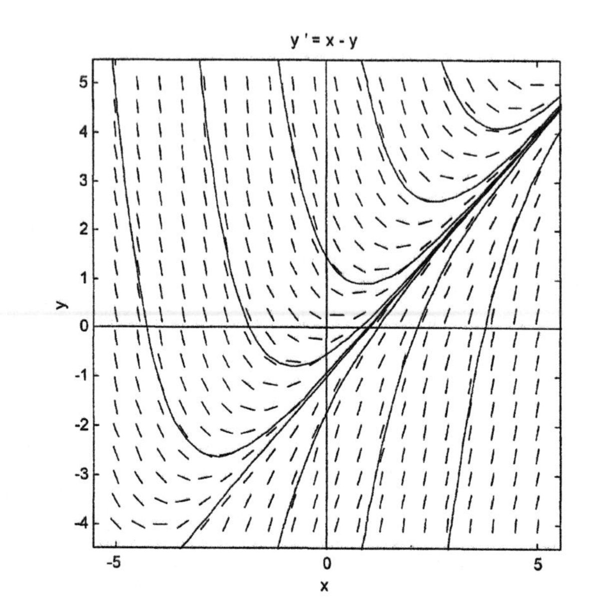

curves approach as $x \to \infty$. Indeed, if we substitute the trial straight line solution $y(x) = ax + b$ in the differential equation, we get

$$a = y' = x - y = x - (ax + b) = (1 - a)x - b,$$

which is so if and only if $a = 1$ and $b = -1$. Thus $y(x) = x - 1$ is, indeed, a straight line solution of the differential equation $y' = x - y$. The figure then suggests (without proving it) that

$$y(x) - (x - 1) \to 0 \quad \text{as} \quad x \to \infty.$$

The following two investigations involve similar inferences from slope fields and computer-generated solution curves. You might warm up by generating the direction fields and some solution curves for Problems 1–10 in this section.

Investigation A
Plot a direction field and typical solution curves for the differential equation $dy/dx = \sin(x - y)$, but with a larger window than that of Fig. 1.3.4. With $-10 \le x, y \le 10$, for instance, a number of apparent straight line solution curves should be visible.

(a) Substitute $y = ax + b$ in the differential equation to determine what the coefficients a and b must be in order to get a solution.

(b) A computer algebra system gives the general solution

$$y(x) = x - 2\tan^{-1}\left(\frac{x - 2 - C}{x - C}\right).$$

Can you determine a value of the arbitrary constant C that yields the linear solution $y = x - \pi/2$ corresponding to the initial condition $y(\pi/2) = 0$?

Investigation B

For your own personal differential equation, let n be the *smallest* integer in your student ID number that is *greater* than 1, and consider the differential equation

$$\frac{dy}{dx} = \frac{1}{n}\cos(x - n\,y).$$

(a) First investigate (as in (a) above) the possibility of straight line solutions.

(b) Then generate a slope field for this differential equation, with the viewing window chosen so that you can picture some of these straight lines, plus a sufficient number of nonlinear solution curves that you can formulate a conjecture about what happens to $y(x)$ as $x \to \infty$. State your inference as plainly as you can. Given the initial value $y(0) = y_0$, it would be nice to be able to say (perhaps in terms of y_0) how $y(x)$ behaves as $x \to \infty$.

(c) A computer algebra system gives the general solution

$$y(x) = \frac{1}{n}\left[x + 2\tan^{-1}\left(\frac{1}{x - C}\right)\right]$$

Can you make a connection between this symbolic solution and your graphically generated solution curves (straight lines or otherwise)?

In the sections that follow we outline the use of *Maple*, *Mathematica*, and MATLAB to generate slope fields and solution curves.

Using *Maple*

The differential equation $dy/dx = x - y$ is defined in *Maple* by the command

```
de := diff( y(x), x)  = x - y(x);
```

$$de := \frac{\partial}{\partial x}y(x) = x - y(x)$$

Note that we write $y(x)$, not just y, to specify that the dependent variable y is a function of the independent variable x, and that $:=$ is used as the assignment operator. The differential equation solver in *Maple* is **dsolve**, and it gives the general solution

```
yg := dsolve( de, y(x) );
```

$$yg := \ y(x) = x - 1 + e^{(-x)}\_C1$$

Observe how *Maple* writes the (first) arbitrary constant $\_C1$ that appears. We can verify that the righthand side expression in the equation yg actually satisfies the differential equation by showing that it yields an identity upon substitution.

```
subs( y(x) = rhs(yg), de );
```

$$\frac{\partial}{\partial x}(x - 1 + e^{(-x)}\_C1) \ = \ 1 - e^{(-x)}\_C1$$

When we enter the command **`eval(%)`** to ask that the derivative on the left be evaluated, the identity

$$1 - e^{(-x)}\_C1 \ = \ 1 - e^{(-x)}\_C1$$

that results shows that the general solution of the differential equation **`de`** is, indeed, $y(x) = x - 1 + Ce^{x}$. We can get a particular solution satisfying a given initial condition either by specifying it in the **`dsolve`** command,

```
y1 := dsolve( {de, y(0)=2}, y(x) );
```

$$soln1 := \ y(x) = x - 1 + 3e^{(-x)}$$

or by substituting a specific numerical value for the arbitrary constant in the general solution:

```
y2 := subs(_C1=0, yg );
```

$$soln2 := \ y(x) = x - 1$$

We can plot both these solution curves simultaneously:

```
y1 := rhs(y1):
y2 := rhs(y2):
plot( {y1,y2}, x=-1..3 );
```

The result is shown at the top of page 5.

Any finite number of particular solutions could be plotted simultaneously (as shown in the figure at the top of page 5) by entering them as a set enclosed by braces. However, the **`DEtools`** package contains the special command **`DEplot`** for doing this sort of thing. The following rather detailed command shows how to plot simultaneously a slope field and a collection of solution curves satisfying a set of separate initial conditions, each corresponding to a different initial point (x_0, y_0).

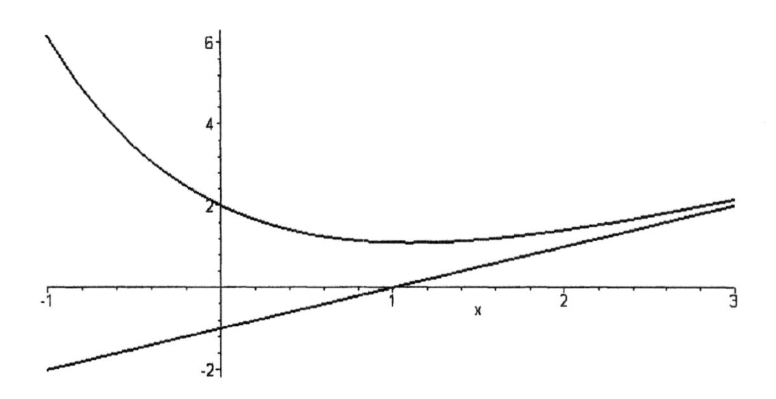


```
with(DEtools):
DEplot( diff(y(x),x) = x - y(x),y(x),
     x=-5..5,  y=-5..5,
     {[0,4],[0,2],[0,1],[0,0],[0 ,-1],[0,-2],
      [0,-4],[-4,-2],[-3,4],[-4,4],[1,4],[2,4],
      [3,4],[-2,-4],[1,-4],[2,-4],[3,-4]},
      dirgrid=[12,12],
      color = black, linecolor = blue,
      thickness = 2 );
```

We have specified the right hand side $f(x, y) = x - y$ of the differential equation $y' = f(x, y)$, the axes $[x, y]$ for plotting, and the window to be plotted.

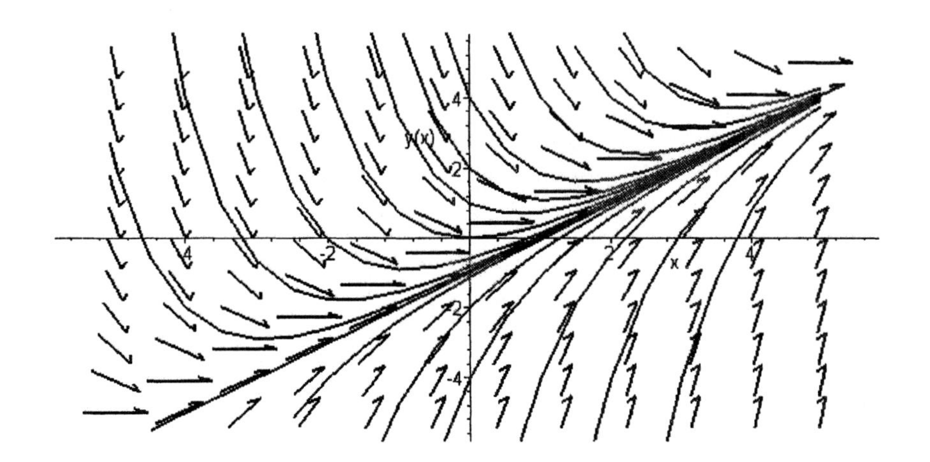

Here we see a pleasant variety of solution curves — all appearing to funnel in on the single linear "asymptotic solution" $y = x + 1$ — together with a slope field consisting of a 12-by-12 **grid** of arrows. We could plot the slope field alone by deleting the initial points.

Using *Mathematica*

To define the differential equation $y' = x - y$ in *Mathematica* we enter the command

```
de = D[y[x],x] == x - y[x]
```

$$y'(x) == x - y(x)$$

Note that *Mathematica* uses = for assignment, == for ordinary equality, and square brackets for functional notation. The differential equation solver in *Mathematica* is **DSolve**.

```
soln = DSolve[ de, y[x], x ]
```

$$\{\{y(x) \rightarrow x + e^{-x} c_1 - 1\}\}$$

We can define the general solution explicitly as a function of **x** by taking the 2nd element of the 1st element of the 1st element of the nested list **soln**:

```
yg[x_] = soln[[1,1,2]]
```

$$x + e^{-x} c_1 - 1$$

We can verify this general solution by substituting it for the function **y** in the differential equation.

```
de /. y -> yg
```

True

We can get a particular solution satisfying a given initial condition by specifying it in the **DSolve** command:

```
y1 = DSolve[{de, y[0] == 2}, y[x], x];
y1 = y1[[1,1,2]] // Simplify
```

$$x + 3e^{-x} - 1$$

Alternatively, we can substitute a numerical value for the arbitrary constant in the general solution:

```
y2 = yg[x] /. C[1] -> 0
```

$$x - 1$$

We can plot both these solution curves simultaneously; the command

```
Plot[ {y1,y2}, {x, -1,3} ];
```

produces the same two-curve figure generated previously using *Maple*. Indeed, any finite number of particular solutions could be plotted simultaneously by entering them as a list enclosed by braces. We can construct just such a list by substituting a list (or table) of numerical values for the arbitrary constant C[1] = c_1 in our general solution.

```
yp = yg[x] /. C[1] -> Table[c, {c,-5,5} ]
```

$$\{x-5e^{-x}-1,\ x-4e^{-x}-1,\ x-3e^{-x}-1,\ x-2e^{-x}-1,\ x-e^{-x}-1,$$
$$x-1,\ x+e^{-x}-1,\ x+2e^{-x}-1,\ x+3e^{-x}-1,\ x+4e^{-x}-1,\ x+5e^{-x}-1\}$$

Here we have specified the values `C[1] = -5,-4, ....., 4,5.`

```
curves =
Plot[ Evaluate[yp], {x,-5,5}, PlotRange -> {-5,5}];
```

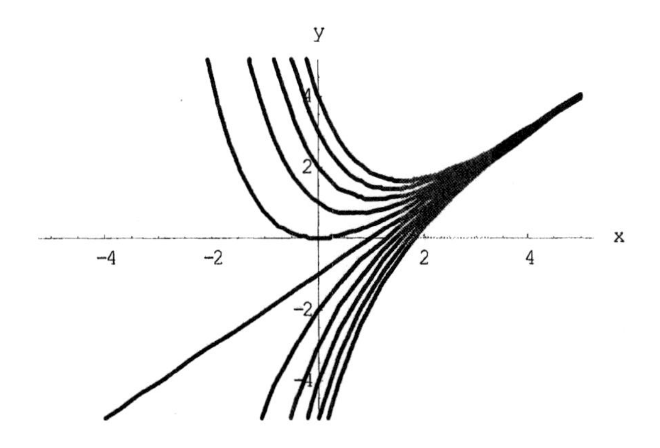

Here we see a family of solution curves, all appearing to funnel in on the single "asymptotic solution" $y = x + 1$. We can use the **PlotVectorField** function, after loading the *Mathematica* package

```
Needs["Graphics`PlotField`"]
```

to superimpose a slope field. Then the commands

```
slopes =
PlotVectorField[{1,x-y}, {x,-5,5}, {y,-5,5} ];

Show[ curves, slopes, AspectRatio -> 1 ];
```

do the job; try it for yourself.

Using MATLAB

Figures including slope fields are generated with the greatest ease using the menu-driven **dfield** function in the *ODE Using MATLAB* package mentioned previously. Here we illustrate also the "hands on" generation of symbolic solutions and solution curves using the Student Edition of MATLAB (version 5) — or the professional edition equipped with the Symbolic Math Toolbox, which calls on an imbedded *Maple* kernel for the execution of symbolic operations.

Symbolic variables, formulas, or equations are entered as strings enclosed in single quotes. Thus we can define the differential equation $y' = x - y$ by entering

```
syms x y
de = 'Dy = x - y'

de =
Dy = x - y
```

with **D** denoting differentiation of the dependent variable **y** (which immediately follows the **D**) with respect to the independent variable **x** (the other variable in the equation). The function **dsolve** computes explicit symbolic solutions (when possible). Thus:

```
yg = dsolve(de,'x');
yg = expand(yg)

yg =
x-1+1/exp(x)*C1
```

Apparently we have a general solution involving an arbitrary constant C1. To verify this, we check that when we differentiate the expression **yg** with respect to the symbolic variable **x**,

```
Dy = diff(yg,x)

Dy =
1-1/exp(x)*C1
```

We get the same result as when we substitute **yg** for **y** in the differential equation,

```
subs(de, y, yg)

ans =
Dy = 1-1/exp(x)*C1
```

[The syntax for symbolic substitution of **new** for **old** in the expression **expr** is **subs(expr, old, new)**.]

We can get a particular solution satisfying a given initial condition either by specifying it in the **dsolve** command,

```
y1 = dsolve(de, 'y(0)=2', 'x');
y1 = expand(y1)

y1 =
x-1+3/exp(x)
```

or by substituting a specific numerical value for the arbitrary constant in the general solution:

```
y2 = subs(yg,'C1',0)

y2 =
x-1
```

We can use MATLAB's **ezplot** command to plot both these two solution curves with the commands

```
ezplot(y1,[-1 3])
hold on
ezplot(y2,[-1 3])
axis([-1 3 -2 6])
grid on
```

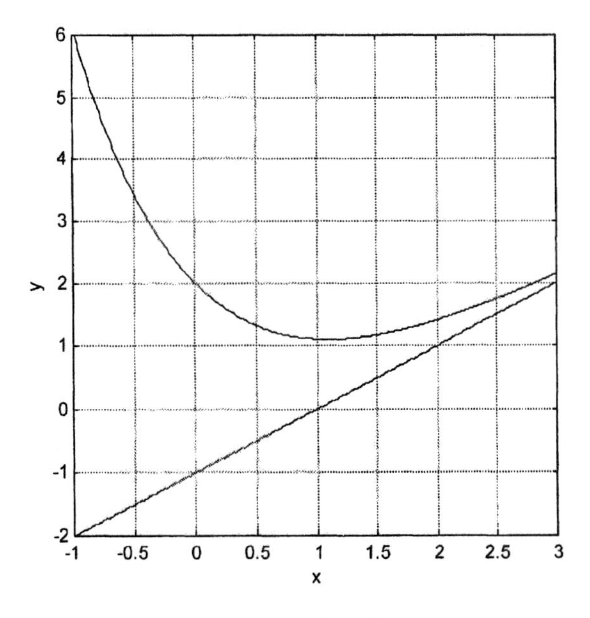

The command **ezplot('f', [a b])** plots the expression **'f'** as a function of **x** on the interval $[a, b]$. The **hold on** command holds the first solution curve in place while the while the second one is plotted. The **axis** command overrides MATLAB's auto-sizing and shows the picture in the desired viewing window:

Unless the special purpose **dfield** program is used, plotting slope fields in MATLAB requires a bit of work. The following program utilizes the built-in command **quiver(x,y,dx,dy)** that plots an $n \times n$ array of vectors with x- and y- components specified by the $n \times n$ matrices **dx** and **dy**, based at the xy-points in the plane whose x- and y- coordinates are specified by the $n \times n$ matrices **x** and **y**.

```
% slope field program  sfield.m
n = 10;                                % no of subintervals
a = -5; b = 5;                         % x-interval
c = -5; d = 5;                         % y-interval
h = (b-a)/n;   k = (d-c)/n;            % x- and y-step sizes
x = a : h : b;                         % x-subdivision points
y = c : k : d;                         % y-subdivision points
[x,y] = meshgrid(x,y);                 % grid of points
f = x - y;                             % define f(x,y)
t = atan(f);                           % angle of inclination
dx = cos(t);     dy = sin(t);          % xy-components of arrow
quiver(x,y,dx,dy)                      % plot slope field
axis([a b c d])                        % viewing window
grid on                                % draw grid lines
```

However, you need not be concerned about these matrices of coordinates and components — only the number n of subintervals in each direction, the desired viewing window $a \leq b,\ c \leq d,$ and the expression **f** defining the right-hand side of the differential equation $y' = f(x,y)$ need be altered when the program **sfield** is defined by saving the commands listed above in the text file **sfield.m**. (Alternatively, these commands could be entered individually in command mode). For instance, once the slope field program **sfield** has been defined, the commands

```
for k = -5 : 5
     ezplot(subs(yg,'C1',k),[-5 5]);
     hold on
     end
sfield
```

first plot the particular solution curves corresponding to the values –5, –4, ..., 4, 5 of the arbitrary constant **C1** in the general solution **yg**, and then a 10×10 grid of direction field arrows is added. The result is shown on the next page.

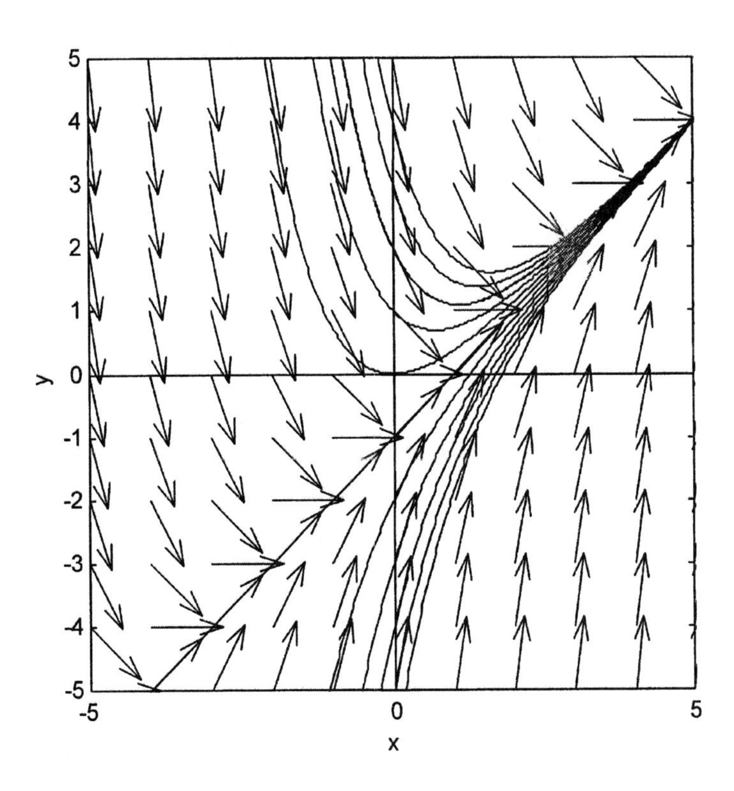

5
4
3
2
1
0
-1
-2
-3
-4
-5
y
-5
0
5
x

1.4 Application
Separable Equations and the Logistic Equation

If a separable differential equation is written in the form $f(y)\,dy = g(x)\,dx$, then its general solution can be written in the form

$$\int f(y)\,dy = \int g(x)\,dx + C.$$

Thus the solution of a separable differential equation reduces to the evaluation of two indefinite integrals. Hence it is tempting to use a computer algebra system such as *Maple* or *Mathematica* that can compute such integrals symbolically.

We illustrate this approach using the *logistic differential equation*

$$\frac{dx}{dt} = ax - bx^2 \tag{1}$$

that models a population $x(t)$ with births (per unit of time) proportional to x and deaths proportional to x^2. If $a = 0.01$ and $b = 0.0001$, for instance, Eq. (1) is

$$\frac{dx}{dt} = 0.01x - 0.0001x^2 = \frac{x}{10000}(100 - x). \tag{2}$$

Separation of variables leads to

$$\int \frac{dx}{x(100-x)} = \int \frac{dt}{10000} = \frac{t}{10000} + C. \tag{3}$$

Any computer algebra system gives a result of the form

$$\frac{1}{100}\ln(x) - \frac{1}{100}(x - 100) = \frac{t}{10000} + C. \tag{4}$$

You can now apply the initial condition $x(0) = x_0$, combine logarithms, and finally exponentiate in order to solve (4) for the particular solution

$$x(t) = \frac{100\,x_0 e^{t/100}}{100 - x_0 + x_0 e^{t/100}} \tag{5}$$

of (2). The direction field and solution curves shown in Fig. 1.4.13 in the text suggest that, whatever is the initial value x_0, the solution $x(t) \to 100$ as $t \to \infty$. Can you use (5) to verify this conjecture?

The sections that follow illustrate the use of *Maple*, *Mathematica*, and MATLAB to carry out the procedure outlined above. You might warm up for the investigation below by applying a computer algebra system to solve Problems 1–28 in Section 1.4 of the text.

Investigation
For your own personal logistic equation, take $a = m/n$ and $b = 1/n$ in (1), with m and n being the *largest* two distinct digits (in either order) in you student ID number.

(i) First generate a slope field for your differential equation and include a sufficient number of solution curves that you can see what happens to the population as $t \to \infty$. State your inference plainly.

(ii) Next, use a computer algebra system to solve the differential equation symbolically, and use the symbolic solution to find the limit of $x(t)$ as $t \to \infty$. Was your graphically-based inference correct?

(iii) Finally, state and solve a numerical problem using the symbolic solution. For instance, how long does it take x to grow from a selected initial value x_0 to a given target value x_1?

Using *Maple*

First we integrate both sides of our separated differential equation as in Eq. (3).

```
soln := int(1/(x*(100-x)),x) = int(1/10000,t)+C;
```

$$soln := \frac{1}{100}\ln(x) - \frac{1}{100}\ln(-100+x) = \text{int}(1/10000, t) + C$$

Then we apply the initial condition $x(0) = x0$ to find the constant C.

```
C := solve(subs(x=x0,t=0,soln),C);
```

$$C := \frac{1}{100}\ln(x0) - \frac{1}{100}\ln(-100+x0)$$

We substitute this value of C and simplify.

```
soln := simplify(100*soln);
```

$$soln := \ln(x) - \ln(-100+x) = \frac{1}{100}t + \ln(x0) - \ln(-100+x0)$$

Next we exponentiate both sides of this equation.

```
soln := simplify(exp(lhs(%)) = exp(rhs(%)));
```

$$soln := \frac{x}{-100+x} = \frac{e^{\left(\frac{1}{100}t\right)}x0}{-100+x0}$$

Finally we solve explicitly for x as a function of t,

```
x(t) = solve(soln, x);
```

$$x(t) = 100\frac{e^{\left(\frac{1}{100}t\right)}x0}{100-x0+e^{\left(\frac{1}{100}t\right)}x0}$$

as in Eq. (5) above.

Using *Mathematica*

First we integrate both sides of our separated differential equation as in Eq. (3).

```
soln =
Integrate[1/(x(100-x)),x] == Integrate[1/10000,t] + c
```

$$\frac{\log(x)}{100} - \frac{1}{100}\log(x-100) == c + \frac{t}{10000}$$

Then we apply the initial condition $x(0) = x0$ to find the constant c.

```
c = First[ soln /. {t->0, x->x0} ]
```

$$\frac{\log(x0)}{100} - \frac{1}{100}\log(x0-100)$$

We substitute this value of c and simplify.

```
soln =
Expand[100*First[soln]] == Expand[100*Last[soln]]
```

$$\log(x) - \log(x-100) == \frac{t}{100} - \log(x0-100) + \frac{1}{100}\log(x0)$$

Next we exponentiate both sides of this equation.

```
soln =
```

```
Exp[First[soln]] == Exp[Last[soln]] // Simplify
```

$$\frac{x}{x-100} == \frac{e^{t/100}x0}{x0-100}$$

Finally we solve explicitly for x as a function of t,

```
soln = Solve[ soln, x ];
```

$$\left\{\left\{x \to \frac{100e^{t/100}x0}{e^{t/100}x0 - x0 + 100}\right\}\right\}$$

```
x = First[x /. soln]
```

$$\frac{100e^{t/100}x0}{e^{t/100}x0 - x0 + 100}$$

as in Eq. (5) above.

Using MATLAB

Here we solve the logistic equation in (2) using the MATLAB "symbolic toolbox" interface to the *Maple* kernel. We begin by separating variables and integrating each side of the resulting equation. However, it is more convenient now to work with "everything on one side of the equation", as in

$$\int \frac{dx}{x(100-x)} - \int \frac{dt}{10000} - C = 0.$$

So we start by "declaring" our symbolic variables and evaluating the two integrals in this equation.

```
syms x t C
soln = int(1/(x*(100-x)),x) - int(1/10000, t) - C
```

```
soln =
1/100*log(x)-1/100*log(-100+x)-1/10000*t-C
```

We are actually thinking here of the equation `soln = 0`, but the right-hand side zero is suppressed throughout. It simplifies the equation a bit by multiplying through by 100.

```
soln = 100*soln
```

```
soln =
log(x)-log(-100+x)-1/100*t-100*C
```

Then we apply the initial condition $x(0) = x0$ to find the constant C.

```
soln0 = subs(soln, {t,x}, {0,'x0'})

soln0 =
log(x0)-log(-100+x0)-100*C

C = solve(soln0, C)

C =
1/100*log(x0)-1/100*log(-100+x0)
```

We substitute this value of C simply by evaluating the present implicit solution.

```
soln = eval(soln)

soln =
log(x)-log(-100+x)-1/100*t-log(x0)+log(-100+x0)
```

Finally we solve explicitly for x as a function of t,

```
x = solve(soln, x);
pretty(x)

              x0 exp(1/100 t)
100 --------------------------
      100 - x0 + x0 exp(1/100 t)
```

as in Eq. (5) above.

Further Investigation

Let's return to the general solution $\int f(y)dy = \int g(x)dx + C$ of the separable differential equation $f(y)dy = g(x)dx$. Assuming that we can find both of the indefinite integrals $F(y) = \int f(y)dy$ and $G(x) = \int g(x)dx$, we can write the general solution of the equation as $H(x,y) = C$, where $H(x,y)$ denotes $F(y) - G(x)$. Now suppose we were to *graph* $H(x,y)$ in three dimensions as a function of x and y; could we tell somehow from this graph that the original differential equation was separable? We illustrate this investigation using *Mathematica*; *Maple* and MATLAB users can substitute equivalent commands.

We'll try this on an example that only a computer algebra system could love:

$$\sin\left(\frac{y}{3}\right)\cos(2y)e^{y/10}\,dy = \sin^7 x\,dx$$

(If we *were* to solve this equation by hand, then what integration techniques would be called for?) The following *Mathematica* commands will quickly find $H(x,y)$ for us:

```
lhs = Sin[y/3]*Cos[2y]*Exp[y/10]
rhs = (Sin[x])^7
H[x_, y_] = Integrate[lhs, y] - Integrate[rhs,x]
```

Then this command will plot $H(x,y)$ over $-4 \le x, y \le 4$:

```
Plot3D[ H[x, y], {x, -4, 4}, {y, -4, 4},
     ViewPoint->{3, 1.25, 1.2}, BoxRatios->{1, 1, 1}];
```

Is there anything about this graph so far that would shout "separable equation"?

Let's dig further by looking at *cross-sections* of this graph perpendicular to the x-axis. The following command will generate frames of an animation, and then double-click on any of the frames to play the animation (during play you can use the number keys 1,2,3,... to adjust the speed of play):

```
Table[ ParametricPlot3D[{x, t, H[x, t]}, {t, -4, 4},
       PlotRange -> {{-4, 4}, {-4, 4}, {-1.25, 1.25}},
       ViewPoint->{3, 1.25, 1.2},
       BoxRatios -> {1, 1, 1}],
       {x, -4, 4, 0.5}];
```

What if $H(x,y)$ had been a *randomly chosen* function of two variables; do you think that the cross-sections of the graph of H would have all had the same shape, and just bounced around, as these do?

Similarly we can look at the cross-sections perpendicular to the y-axis:

```
Table[ ParametricPlot3D[{t, y, H[t, y]}, {t, -4, 4},
       PlotRange -> {{-4, 4}, {-4, 4}, {-1.25, 1.25}},
       ViewPoint -> {3, 1.25, 1.2},
       BoxRatios -> {1, 1, 1}], {y, -4, 4, 0.5}];
```

Do these curves show the same general behavior?

By modifying the above commands, experiment with separable equations of your own devising. Do the cross sections always behave in this way? Can you explain why?

1.5 Application
Linear Equations and Temperature Oscillations

Section 1.5 of the text describes the following 4-step algorithm for solving the linear first-order differential equation

$$\frac{dy}{dx} + p(x)\,y = q(x). \tag{1}$$

1. Begin by calculating the *integrating factor*

$$\rho(x) = e^{\int p(x)\,dx}. \tag{2}$$

2. Then multiply both sides of the differential equation by $\rho(x)$.

3. Next, recognize the left-hand side of the resulting equation as the derivative of a product, so it takes the form

$$D_x[\rho(x)\,y(x)] = \rho(x)\,q(x). \tag{3}$$

4. Finally, integrate this last equation to get

$$\rho(x)\,y(x) = \int \rho(x)\,q(x)\,dx + C, \tag{4}$$

and then solve for $y(x)$ to obtain the general solution of the original differential equation in (1).

This algorithm is well-adapted to automatic symbolic computation. In the technology-specific sections of this project, we illustrate its implementation using *Maple*, *Mathematica*, and MATLAB to solve the initial value problem

$$\frac{dy}{dx} - 3\,y = e^{2x}, \qquad y(0) = 3. \tag{5}$$

In each case, the commands shown here constitute a "template" that you can apply to any given linear first-order differential equation. First redefine the coefficient functions $p(x)$ and $q(x)$ as specified by *your* linear equation, then work through the subsequent steps. You may apply whatever computer algebra system is available to carry out this algorithmic process for the initial value problem in (5), and then apply it to a selection of other examples and problems in Section 1.5 of the text.

For an applied problem to solve in this manner, consider indoor temperature oscillations that are "driven" by outdoor temperature oscillations of the form

$$A(t) = a_0 + a_1 \cos\frac{\pi t}{12} + b_1 \sin\frac{\pi t}{12} \tag{6}$$

having a period of 24 hours (so the cycle of outdoor temperatures repeats itself daily). For instance, for a typical July day in Athens, GA with a minimum temperature of 70°F when $t = 4$ (4 am) and a maximum of 90°F when $t = 16$ (4 pm), we would take

$$A(t) = 80 - 10\cos\omega(t-4) = 80 - 5\cos\omega t - 5\sqrt{3}\sin\omega t \tag{7}$$

with $\omega = \pi/12$, using the trigonometric identity $\cos(\alpha - \beta) = \cos\alpha\cos\beta + \sin\alpha\sin\beta$ to get $a_0 = 80$, $a_1 = 5$, and $b_1 = -5\sqrt{3}$ in (22).

If we write Newton's law of cooling (Eq. (3) of Section 1.1 of the text) for the corresponding indoor temperature $u(t)$ at time t, but with the outside temperature $A(t)$ given by (6) instead of a constant ambient temperature A, we get the linear first-order differential equation

$$\frac{du}{dt} = -k(u - A(t)) \tag{8}$$

with coefficient functions $P(t) = k$ and $Q(t) = k\,A(t)$. Typical values of the proportionality constant k range from 0.2 to 0.5 (though k might be greater than 0.5 for a poorly insulated building with open windows, or less than 0.2 for a well-insulated building with tightly sealed windows).

Scenario

Suppose our air conditioner fails at time $t_0 = 0$ one midnight, and we cannot afford to get it repaired until payday at the end of the month. We therefore want to investigate the resulting indoor temperatures that we must endure for the next several days.

Begin your investigation by using a computer algebra system to solve Eq. (8) with $A(t)$ given by (7) and with the initial condition $u(0) = u_0$ (the indoor temperature at the time of the air conditioner failure). (Remember that the independent and dependent variables are t and u instead of x and y.) You should get the solution

$$u(t) = 80 + c_0 e^{-kt} + c_1\cos\omega t + d_1\sin\omega t \tag{9}$$

where

$$c_0 = u_0 - \frac{75k^2 + 5\sqrt{3}\,k\omega + 80\omega^2}{k^2 + \omega^2},$$

$$c_1 = \frac{-5k^2 + 5\sqrt{3}\,k\omega}{k^2 + \omega^2}, \qquad d_1 = \frac{-5\sqrt{3}\,k^2 - 5k\omega}{k^2 + \omega^2}$$

with $\omega = \pi/12$. With $k = 0.2$ (for instance), this solution reduces (approximately) to

$$u(t) = 80 + e^{-t/5}(u_0 - 82.3351) + 2.3351\cos\frac{\pi t}{12} - 5.6036\sin\frac{\pi t}{12}. \qquad (10)$$

Observe first that the damped exponential term in (10) approaches 0 as $t \to \infty$, leaving the long-term "steady periodic" solution

$$u_{sp}(t) = 80 + 2.3351\cos\frac{\pi t}{12} - 5.6036\sin\frac{\pi t}{12}. \qquad (11)$$

Consequently, the long-term indoor temperatures oscillate every 24 hours about the same average temperature of 80°F as the outdoor average temperature.

Fig. 1.5.10 in the text shows a number of solution curves corresponding to possible initial u_0 temperatures ranging from 65°F to 95°F. Observe that — whatever the initial temperature — the indoor temperature "settles down" within about 18 hours to a periodic daily oscillation. However, the "amplitude" of temperature variation is less indoors than outdoors. Indeed, using the trigonometric identity mentioned previously, Eq. (11) can be rewritten (verify this!) as

$$u(t) = 80 - 6.0707\cos\left(\frac{\pi t}{12} - 1.9656\right) = 80 - 6.0707\cos\frac{\pi}{12}(t - 7.5082). \qquad (12)$$

Do you see that this implies that the indoor temperature varies between a minimum of about 74°F and a maximum of about 86°F?

Finally, comparison of (7) and (12) indicates that the indoor temperature lags behind the outdoor temperature by about $7.5082 - 4 \approx 3.5$ hours, as illustrated in Fig. 1.5.11. Thus the temperature inside the house continues to rise until about 7:30 pm each evening, so the hottest part of the day inside is early evening, rather than late afternoon (as outside).

For a personal problem to investigate, carry out a similar analysis using average July daily maximum/minimum figures for your own locale, and a value of k appropriate to your own home. You might also consider a winter day instead of a summer day. (What is the winter-summer difference for the indoor temperatures problem?)

Using *Maple*

We first define the coefficient functions and initial values for the initial value problem specified in (5).

```
p := -3:
q := exp(2*x):
```

```
x0 := 0:
y0 := 3:
```

Then we calculate the integrating factor defined in (2),

```
rho := exp( int(p,x) );
```

$$\rho := e^{(-3x)}$$

Equation (4) above now gives the general solution

```
yg := (1/rho)*(c + int(rho*q, x));
```

$$yg := \frac{c - \frac{1}{e^x}}{e^{(-3x)}}$$

```
yg := simplify(%);
```

$$yg := e^{(3x)}(c - e^{(-x)})$$

If we want a particular solution, we need only substitute the given initial values and solve for the constant c.

```
c := eval( solve(subs(x=x0,yg)=y0,c) );
```

$$c := 4$$

Now that $c = 4$, we need only evaluate our general solution to get the desired particular solution.

```
yp := eval(yg);
```

$$yp := e^{(3x)}(4 - e^{(-x)})$$

Frequently it is necessary to expand and simplify an expression to get the "best" form.

```
yp := simplify(expand( yp ));
```

$$yp := 4e^{(3x)} - e^{(2x)}$$

Using *Mathematica*

We first define the coefficient functions and initial values for the initial value problem specified in (5).

```
p = -3;
q =  Exp[2x];
x0 = 0
y0 = 3;
```

Then we calculate the integrating factor defined in (2),

```
rho = Exp[ Integrate[p, x] ]
```

e^{-3x}

Equation (4) above now gives the general solution

```
yg = (1/rho)(c + Integrate[rho*q, x])
```

$e^{3x}\left(c - e^{-x}\right)$

If we want a particular solution, we need only substitute the given initial values and solve for the constant c.

```
Solve[(yg /. x -> x0) == y0, c];
c = c /. First[%]
```

4

Now that $c = 4$, we need only evaluate our general solution to get the desired particular solution

```
yp = yg // Expand
```

$-e^{2x} + 4e^{3x}$

Frequently it is necessary to expand and/or simplify an expression to get the "best" form.

Using MATLAB

We first define the coefficient functions and initial values for the initial value problem specified in (5).

```
syms x y c
p = -3;
q = exp(2*x);
x0 = 0;
y0 = 3;
```

Then we calculate the integrating factor defined in (2),

```
rho = exp(int(p,x))

rho =
exp(-3*x)
```

Equation (4) above now gives the general solution

```
yg = (1/rho)*(c + int(rho*q, x))

yg =
1/exp(-3*x)*(c-1/exp(x))
```

That is, $y(x) = e^{3x}(c - e^{-x})$. To find the desired particular solution, we substitute $x = x_0$ and solve the equation $y(x) = x_0$ for the numerical value of the arbitrary constant c.

```
solve( subs(yg, x,x0)-y0, c)

ans =
4
```

Finally, we get the desired particular solution by substituting $c = 4$ in the general solution.

```
yp = expand(subs(yg, c,4))

yp =
4*exp(x)^3-exp(x)^2
```

That is, $y(x) = 4e^{3x} - e^{2x}$.

1.6 Application
Computer Algebra Solutions

Computer algebra systems typically include commands for the "automatic" solution of differential equations. However, two different systems often give different results whose equivalence is not clear, and a single system may give the solution in an overly complicated form. Consequently, computer algebra solutions of differential equations often require considerable "processing" or simplification by a human user in order to be yield concrete and applicable information. Here we illustrate these issues using the interesting differential equation

$$\frac{dy}{dx} = \sin(x-y) \tag{1}$$

that appeared in the Section 1.3 Application. The *Maple* command

```
dsolve( D(y)(x) = sin(x - y(x)), y(x));
```

yields the nice-looking result

$$y(x) = x - 2\tan^{-1}\left(\frac{x-2-C1}{x-C1}\right) \tag{2}$$

that was cited there. However, the supposedly equivalent *Mathematica* command

```
DSolve[ y'[x] == Sin[x - y[x]], y[x], x]
```

returns the decidedly more complicated-looking result

$$y(x) = x \pm 2\cos^{-1}\left(\pm\frac{\sqrt{4+4x+x^2+4C1+2xC1+C1^2}}{\sqrt{4+4x+2x^2+4C1+4xC1+2C1^2}}\right) \tag{3}$$

This apparent disparity is not unusual, and with another differential equation it might be the other way around. As an alternative to attempting to reconcile the results in (2) and (3), a common tactic is to simplify the differential equation in advance, before submitting it to a computer algebra system. Exercises 1 through 5 below outline a preliminary paper-and-pencil investigation of Eq. (1).

Exercise 1 Show that the plausible substitution $v = x - y$ in (1) yields the separable equation

$$\frac{dv}{dx} = 1 - \sin v. \tag{4}$$

Now the *Mathematica* command

```
Integrate[1/(1 - Sin[v]), v] //  TrigReduce
```

yields

$$\int \frac{dv}{1-\sin v} = \frac{\cos v}{1-\sin v} \qquad \text{(plus a constant)} \tag{5}$$

Exercise 2 Apply the trig identities $\sin 2\alpha = 2\sin\alpha\cos\alpha$ and $\cos 2\alpha = \cos^2\alpha - \sin^2\alpha$ to deduce from (5) that

$$\int \frac{dv}{1-\sin v} = \frac{1+\tan(v/2)}{1-\tan(v/2)} \qquad \text{(plus a constant)} \tag{6}$$

Exercise 3 Deduce from (6) that Eq. (4) has the general solution

$$v(x) = 2\tan^{-1}\left(\frac{x-1+C}{x+1+C}\right)$$

and hence that Eq. (1) has the general solution

$$y(x) = x - 2\tan^{-1}\left(\frac{x-1+C}{x+1+C}\right). \tag{7}$$

Exercise 4 Finally, reconcile the forms in (2) and (7). What is the relation between the arbitrary constants C and $C1$?

Exercise 5 Show that the integral in (5) yields immediately the graphing calculator implicit solution shown in Fig. 1.6.10 of the text.

Investigation
For your own personal differential equation, let p and q be two distinct nonzero integers in your student ID number, and consider the differential equation

$$\frac{dy}{dx} = \frac{1}{p}\cos(x - q\,y). \tag{8}$$

(a) Find a symbolic general solution using a computer algebra system and/or some combination of the techniques illustrated above.

(b) Determine the symbolic particular solution corresponding to several typical initial conditions of the form $y(x_0) = y_0$.

(c) Determine the possible values of a and b such that the straight line $y = ax + b$ is a solution curve of (8).

(d) Plot a direction field and some typical solution curves. Can you make a connection between the symbolic solution and your (linear and nonlinear) solution curves?

The sections that follow illustrate the use of computer algebra systems to implement the substitution techniques of Section 1.6. You try out the technique of computer-algebra substitution with the examples and with Problems 1–30 in Section 1.6.

Using *Maple*

As an example we consider the differential equation

$$\frac{dy}{dx} = (x + y + 3)^2 \tag{9}$$

of Example 1 in the text, which calls for the substitution

$$v = x + y + 3. \tag{10}$$

We first enter our differential equation (9) in terms of y,

```
yDE :=    diff(y(x),x) = (x+y(x)+3)^2;
```

$$yDE := \frac{\partial}{\partial x} y(x) = (x + y(x) + 3)^2$$

and then the substitution

```
vSubst :=    v(x) = x + y(x) + 3;
```

$$vSubst := v(x) = x + y(x) + 3$$

in (10) for v in terms of y. The inverse substitution for y in terms of v will be

```
ySubst := y(x) = solve(vSubst, y(x));
```

$$ySubst := y(x) = v(x) - x - 3$$

After calculating the derivative dy/dx in terms of dv/dx,

```
DySubst := diff(ySubst,x);
```

$$DySubst := \frac{\partial}{\partial x} y(x) = \left(\frac{\partial}{\partial x} v(x)\right) - 1$$

we substitute for $y(x)$ and $y'(x)$ in the original y-equation to get our transformed differential equation

```
vDE := eval(subs(ySubst, DySubst, yDE));
```

$$vDE := \left(\frac{\partial}{\partial x} v(x)\right) - 1 = v(x)^2$$

to be solved for $v(x)$. Having gotten this far with the computer, we might as well call on *Maple*'s **dsolve** function to solve this differential equation.

```
vSolution := dsolve( vDE, v(x) );
```

$$vSolution := v(x) = \tan(x + \_C1)$$

It remains only to re-substitute for v in terms and y and solve explicitly for the solution $y(x)$ of our original differential equation.

```
ySolution := subs(vSubst, vSolution );
```

$$ySolution := x + y(x) + 3 = \tan(x + \_C1)$$

```
y(x) = solve( y_solution, y(x) );
```

$$y(x) = -x - 3 + \tan(x + \_C1)$$

The *Maple* commands shown above provide a "template" that can be used with more complicated differential equations. You need only re-enter your own differential equation **yDE** and your desired substitution **vSubst**, and then proceed to re-execute the remaining commands.

Using *Mathematica*

Here we illustrate the use of *Mathematica* to solve a homogeneous differential equation of the form

$$\frac{dy}{dx} = F\left(\frac{y}{x}\right) \tag{11}$$

using the standard substitutions

$$v = \frac{y}{x}, \quad y = vx, \qquad \frac{dy}{dx} = v + x\frac{dv}{dx}. \tag{12}$$

To solve the differential equation

$$2xy\frac{dy}{dx} = 4x^2 + 3y^2 \tag{13}$$

of Example 2 in the text, for instance, we first enter the equation as

```
yDE  =  2 x y[x] y'[x] == 4 x^2 + 3 y[x]^2;
```

The substitution

```
y[x_] := x v[x]
```

in (12) then yields the transformed differential equation

```
vDE = yDE
```

$$2x^2v(x)(v(x) + x\,v'(x)) == 3\,v(x)^2x^2 + 4x^2$$

when we re-evaluate **yDE**. Since we're prodeeding entirely by computer, let's just use *Mathematica*'s **DSolve** function to solve for $v(x)$.

```
vSolution = DSolve[vDE, v[x] ,x ]
```

$$\left\{\left\{v(x) \to -\sqrt{x}\sqrt{c_1 - \frac{4}{x}}\right\}, \left\{v(x) \to \sqrt{x}\sqrt{c_1 - \frac{4}{x}}\right\}\right\}$$

```
v[x_] = v[x] /. vSolution
```

$$\left\{-\sqrt{x}\sqrt{c_1 - \frac{4}{x}}, \sqrt{x}\sqrt{c_1 - \frac{4}{x}}\right\}$$

We get two distinct v-solutions differing by sign, and the corresponding y-solutions result when we re-evaluate $y(x)$, thereby automatically substituting each formula for $v(x)$ via $y(x) = x\,v(x)$.

```
ySolution = y[x]
```

$$\left\{-x^{3/2}\sqrt{c_1 - \frac{4}{x}}, x^{3/2}\sqrt{c_1 - \frac{4}{x}}\right\}$$

The implicit solution $y^2 + 4x^2 = Cx^3$ found in the text results when we square both sides.

```
implicitSolution =
y^2 == Expand[Simplify[ First[y[x]]^2 ]] /. C[1] -> C
```

$$y^2 == Cx^3 - 4x^2$$

The *Mathematica* commands shown above provide a "template" that can be used with more complicated differential equations. You need only re-enter your own differential equation **yDE** and your desired substitution — defining **y[x]** in terms of **x** and **v[x]** — and then proceed to re-execute the remaining commands.

Using MATLAB

As an example we consider the differential equation

$$\frac{dy}{dx} = (x + y + 3)^2 \tag{9}$$

of Example 1 in the text, which calls for the substitution

$$v = x + y + 3. \tag{10}$$

We first enter our differential equation (9) in terms of y,

```
syms x y v
yDE = 'Dy = (x+y+3)^2'

yDE =
Dy = (x+y+3)^2
```

and the substitution

```
vSubst = 'x+y+3'

vSubst =
x+y+3
```

for v in terms of y. To invert this substitution, we set up the equation **v = vSubst** and solve for y in terms of v:

```
ySubst = solve(v-vSubst,y)

ySubst =
v-x-3
```

We calculate the derivative

```
DySubst = 'Dv-1'

DySubst =
Dv-1
```

of y in terms of the derivative of v, and then transform the original differential equation **yDE** by succesively substituting **ySubst** for **y** and **DySubst** for **Dy**.

```
tempDE = subs(yDE, ySubst, y)

tempDE =
Dy = v^2

vDE = subs(tempDE, DySubst, 'Dy')

vDE =
Dv-1 = v^2
```

Now we would like to do **dsolve(vDE,'x')**, but for some reason this gives an error message. Hence we write in the differential equation ourselves to solve for v in terms of x:

```
vSolution = dsolve('Dv-1=v^2','x')

vSolution =
tan(x-C1)
```

Finally, we resubstitute for v in terms of y and solve for $y(x)$:

```
vEquation = subs(v-vSolution,vSubst,v)

vEquation =
x+y+3-tan(x-C1)

y = solve(vEquation, y)

y =
-x-3+tan(x-C1)
```

The MATLAB commands shown above provide a "template" that can be used with more complicated differential equations. You need only re-enter your own differential equation **yDE** and your desired substitution **vSubst**, and then proceed to re-execute the remaining commands.

Chapter 2

Mathematical Models and Numerical Methods

2.1 Application
Logistic Modeling of Population Data

This project deals with the problem of fitting a logistic model to given population data. Thus we want to determine the numerical constants a and b so that the solution $P(t)$ of the initial value problem

$$\frac{dP}{dt} = aP + bP^2, \qquad P(0) = P_0 \tag{1}$$

approximates the given values $P_0, P_1, P_2, \ldots, P_n$ of the population at the times $t_0 = 0, t_1, t_2, \ldots, t_n$. If we rewrite Eq. (1) — the logistic equation with $kM = a$ and $k = -b$ — in the form

$$\frac{1}{P}\frac{dP}{dt} = a + bP, \tag{2}$$

then we see that the points

$$\left(P(t_i), \frac{P'(t_i)}{P(t_i)}\right), \qquad i = 0, 1, 2, \ldots, n$$

should all lie on the straight line with y-intercept a and slope b that is defined by the linear function of P that appears on the right-hand side in Eq. (2).

This observation provides a way to find the values of a and b. If we can determine the approximate values of the derivatives $P_1', P_2', P_3', \cdots$ corresponding to the given population data, then we can proceed with the following agenda:

- First plot the points $(P_1, P_1'/P_1)$, $(P_2, P_2'/P_2)$, $(P_3, P_3'/P_3)$, ... on a sheet of graph paper with horizontal P-axis.

- Then use a ruler to draw a straight line that appears to approximate these points well.

- Finally measure this straight line's y-intercept a and slope b.

But where are we to find the needed values of the derivative $P'(t)$ of the (as yet) unknown function $P(t)$? It is easiest to use the approximation

$$P_i' = \frac{P_{i+1} - P_{i-1}}{t_{i+1} - t_{i-1}} \tag{3}$$

that is suggested by Fig. 2.1.7 in the text. Note that, if the t-values for the given data points are equally distributed with successive differences $h = t_{i+1} - t_{i-1}$, then the quotient on the right-hand side in (3) is a "symmetric difference quotient" of the type

$$\frac{\Delta P}{\Delta t} = \frac{P(t_i + h) - P(t_i - h)}{2h}$$

that is used to define the derivative $P'(t)$, and is "centered" at the point t_i. For instance, if we take $i = 0$ as 1790, then the U.S. population data in Fig. 2.1.8 give

$$P_1' = \frac{P_2 - P_0}{t_2 - t_0} = \frac{7.240 - 3.929}{20} \approx 0.166$$

for the slope at the point (t_1, P_1) corresponding to the year 1800.

Investigation A

Use (3) to verify the slope figures shown in the final column of the table of Fig. 2.1.8 in the text, and then plot the points $(P_1, P_1'/P_1), (P_2, P_2'/P_2), \ldots, (P_{11}, P_{11}'/P_{11})$ indicated by the asterisks in Fig. 2.1.9. If an appropriate graphing calculator, spreadsheet, or computer program is available, use it find the straight line $y = a + bP$ as in (2) that best fits these points. If not, draw your own straight line approximating these points, and then measure its intercept a and slope b as accurately as you can. Next, solve the logistic equation in (1) with these numerical parameters, taking $t = 0$ in the year 1800. Finally, compare the predicted 20th century U.S. population figures with the actual data listed in Fig. 2.1.4 in the text.

Investigation B

Repeat Investigation A, but take $t = 0$ in 1900 and use only 20th century population data. Do you get a better approximation for the U.S. population during the final decades of the 20th century?

Investigation C

Model similarly the world population data shown in Fig. 2.1.10 in the text. The Population Division of the United Nations predicts a world population of 8.177 billion in the year 2025. What do you predict?

The Method of Least Squares

So you think there ought to be a better way than eyeballing the placement of a ruler on a sheet a paper? There is, and it dates back to Gauss. Writing Q for the left-hand side in Eq. (2), we seek the straight line

$$Q = a + bP \tag{4}$$

in the PQ-plane that "best fits" n given data points (P_1, Q_1), (P_2, Q_2), . . . , (P_n, Q_n). Gauss' idea was to choose this line so that it minimizes the *sum of the squares* of the vertical distances between the line and these points. The vertical distance — the difference in Q-coordinates above P_i — between the ith point (P_i, Q_i) is $d_i = Q_i - (a + bP_i)$, so the sum of the squares of these "errors" is the function

$$f(a,b) = \sum_{i=1}^{n} [Q_i - (a + bP_i)]^2 \tag{5}$$

of the (as yet) unknown parameters a and b.

To minimize the value $f(a,b)$ as a function of a and b, we calculate the partial derivatives

$$\frac{\partial f}{\partial a} = \sum_{i=1}^{n} 2[Q_i - (a + bP_i)](-1) = 2a\sum_{i=1}^{n} 1 + 2b\sum_{i=1}^{n} P_i - 2\sum_{i=1}^{n} Q_i$$

and

$$\frac{\partial f}{\partial b} = \sum_{i=1}^{n} 2[Q_i - (a + bP_i)](-P_i) = 2a\sum_{i=1}^{n} P_i + 2b\sum_{i=1}^{n} P_i^2 - 2\sum_{i=1}^{n} P_i Q_i \,.$$

When we set both partial derivatives equal to zero, we get the pair

$$a\,n + b\sum_{i=1}^{n} P_i = \sum_{i=1}^{n} Q_i \tag{6}$$

$$a\sum_{i=1}^{n} P_i + b\sum_{i=1}^{n} P_i^2 = \sum_{i=1}^{n} P_i Q_i \tag{7}$$

of *linear* equations in the unknowns a and b, with coefficients that are simple combinations of the P- and Q-coordinates of the n given data points. It remains only to solve Eqs. (6)–(7) for the coefficients a and b in the desired straight line (4) that best fits these data points.

The sum of squares of errors in (5) is commonly denoted by SSE. Then the **average error** in the linear approximation (4) is defined by

$$\text{average error} = \sqrt{\frac{\text{SSE}}{n}}. \qquad (8)$$

In the following paragraphs we illustrate this least squares approach using *Maple*, *Mathematica*, and MATLAB. In each we consider the $n = 6$ given data points

$$(1.1, 2.88),\ (1.9, 4.36),\ (3.2, 6.98),\ (4.6, 9.86),\ (5.3, 11.20),\ (6.5, 13.54).$$

You can automate similarly your calculations for Investigations A, B, and C.

Using *Maple*

To set up Eqs. (6) and (7) we first enter the lists

```
P := [1.1, 1.9, 3.2, 4.6, 5.3, 6.5]:
Q := [2.88, 4.36, 6.98, 9.86, 11.20, 13.54]:
```

of the P- and Q-coordinates of the

```
n := 6:
```

given data points. Next we use the *Maple* **sum** function to calculate the sums that appear as coefficients of a and b in Eqs. (6)-(7).

```
sumP := evalf(sum( P[i], i = 1..n)):
sumQ := evalf(sum( Q[i], i = 1..n)):
sumPsq := evalf(sum( P[i]^2, i = 1..n)):
sumPQ := evalf(sum( P[i]*Q[i], i = 1..n)):
```

Then the two equations we want to solve are defined by

```
eq6 := n*a + sumP*b = sumQ;
```

$$eq6 := 6\,a + 22.6\,b = 48.82$$

```
eq7 := sumP*a + sumPsq*b = sumPQ;
```

$$eq7 := 22.6\, a + 106.56\, b = 226.514$$

Now we simply solve these two equations for the coefficients a and b using the command

```
soln := fsolve({eq6,eq7}, {a,b});
```

$$\{a = .6457449574, b = 1.988740277\}$$

```
asoln := soln[1]:     bsoln := soln[2]:
a := rhs(asoln);      b := rhs(bsoln);
```

$$a := .6457449574$$
$$b := 1.988740277$$

Thus the least squares best fit in Eq. (4) takes the form

$$Q = 0.6457 + 1.9887\, P$$

(rounding the coefficients to four decimal places). Finally the differences between the actual values $\{Q_i\}$ and the predicted values $\{0.6457 + 1.9887\, P_i\}$ are given by

```
errors := seq(Q[i] - (a + b*P[i]), i = 1..n);
```

$$errors := .0466, -.0644, -.0297, .0660, .0139, -.0326$$

(again rounding to four decimal places). The sum of squares of errors, and then the average error, are given by

```
SSE := sum(errors[i]^2,i = 1..n);
AveError := sqrt(SSE/n);
```

$$AveError := .0462$$

Using *Mathematica*

To set up Eqs. (6) and (7) we first enter the lists

```
P = {1.1, 1.9, 3.2, 4.6, 5.3, 6.5};
Q = {2.88, 4.36, 6.98, 9.86, 11.20, 13.54};
```

of the P- and Q-coordinates of the

```
n = 6;
```

given data points. Next we use the *Mathematica* **Sum** function to calculate the sums that appear as coefficients of a and b in Eqs. (6)-(7).

```
sumP = Sum[ P[[i]], {i, 1,n} ];
sumQ = Sum[ Q[[i]], {i, 1,n} ];
sumPsq = Sum[ P[[i]]^2, {i, 1,n} ];
sumPQ = Sum[ P[[i]]*Q[[i]], {i, 1,n} ];
```

Then the two equations we want to solve are defined by

```
eq6 = n*a + sumP*b == sumQ
```

$6a + 22.6b == 48.82$

```
eq7 = sumP*a + sumPsq*b == sumPQ
```

$22.6\ a + 106.56\ b == 226.514$

Now we simply solve these two equations for the coefficients a and b using the command

```
soln = NSolve[{eq6,eq7}, {a,b}]
```

$\{\{a \to 0.645745, b \to 1.98874\}\}$

```
a = First[a /. soln];
b = First[b /. soln];
```

Thus the least squares best fit in Eq. (4) takes the form

$$Q = 0.6457 + 1.9887\ P$$

(rounding the coefficients to four decimal places). Finally the differences between the actual values $\{Q_i\}$ and the predicted values $\{0.6457 + 1.9887\ P_i\}$ are given by

```
errors = Q - (a + b P)
```

$\{0.0466, 0.0644, -0.0297, 0.0660, 0.0139, -0.0326\}$

(again rounding to four decimal places). The sum of squares of errors, and then the average error, are given by

```
SSE = Sum[errors[[i]]^2, {i,1,n}];
AveError = Sqrt[SSE/n]
```

```
0.0462169
```

Using MATLAB

To set up Eqs. (6) and (7) we first enter the lists

```
P = [1.1, 1.9, 3.2, 4.6, 5.3, 6.5];
Q = [2.88, 4.36, 6.98, 9.86, 11.20, 13.54];
```

of the *P*- and *Q*-coordinates of the

```
n = 6;
```

given data points. Next we use the MATLAB **sum** function to calculate the sums that appear as coefficients of a and b in Eqs. (6)-(7).

```
sumP = sum(P);
sumQ = sum(Q);
sumPsq = sum(P.*P);
sumPQ = sum(P.*Q);
```

Then the two symbolic equations **eq6 = 0** and **eq7 = 0** that we want to solve are defined by

```
eq6 = n*a + sumP*b - sumQ

6*a+113/5*b-2441/50

eq7 = sumP*a + sumPsq*b - sumPQ

113/5*a+2664/25*b-7969752859329691/35184372088832
```

Apparently MATLAB is rationalizing the numerical coefficients. For instance, checking the "big fraction" we see here:

```
sumPQ

sumPQ =
   226.5140

7969752859329691/35184372088832

ans =
   226.5140
```

We proceed to solve symbolically for a and b.

```
soln = solve(eq6,eq7)

soln =
   a: [1x1 sym]
   b: [1x1 sym]
```

This means that **soln** is a structure with two symbolic fields **a** and **b**. The numerical values of these coefficients are given by

```
a = numeric(soln.a)
a =
      0.6457

b = numeric(soln.b)
b =
      1.9887
```

Thus the least squares best fit in Eq. (4) takes the form

$$Q = 0.6457 + 1.9887\, P$$

(with the coefficients rounded to four decimal places). Finally the differences between the actual values $\{Q_i\}$ and the predicted values $\{0.6457 + 1.9887\, P_i\}$ are given by

```
errors = Q - (a + b*P)

errors =
   0.0466   -0.0644   -0.0297    0.0660    0.0139   -0.0326
```

The sum of squares of errors, and then the average error, are given by

```
SSE = sum(errors.^2);
AveError = sqrt(SSE/n)

AveError =
    0.0462
```

Further Investigation

The logistic equation of this applied section is our prime example of a first-order autonomous equation whose solution structure is essentially determined by its "equilibrium solutions" and their stability properties. This appendix suggests a further investigation of equilibrium solutions of such equations, and particularly the use of bifurcation diagrams to portray the way the equilibrium solutions of an autonomous equation $y' = f(x, a)$ may depend on a parameter a appearing in the right-hand side function. We illustrate this investigation using *Mathematica*; *Maple* and MATLAB users can substitute equivalent commands.

We return to the initial value problem in (1),

$$\frac{dP}{dt} = aP + bP^2, \qquad P(0) = P_0, \tag{1}$$

where (as explained in Section 2.1 of the text) $a = \beta_0 - \delta_0$ and $b = -k$. For convenience we will assume that $k = 1$, so that our differential equation becomes

$$\frac{dP}{dt} = aP - P^2, \tag{1'}$$

where the constant a could be positive, negative, or zero. An *equilibrium solution* of (1') is a solution $P(t) \equiv$ constant (and so $\frac{dP}{dt} \equiv 0$) for all t. *How many* equilibrium solutions does this differential equation have, and *where* are they located? (Consider all cases for a, bearing in mind that population can't be negative!)

Because equilibrium solutions are so crucial to understanding differential equations of this type, we illustrate a couple of ways in which they can be seen graphically. First, the following **Do** loop in *Mathematica* will draw movie frames which show solution families of the equation (1') for $P_0 = 0, 0.5, \ldots, 4$, with the constant a ranging from 0 to 4 in increments of 0.25. Double-click on any of the frames to play the animation; during play you can use the number keys 1,2,3,... to adjust the speed of play:

```
Do[  deqn = P'[t]  == a*P[t]  - P[t]^2;
     solnFamily =
          Table[NDSolve[{deqn, P[0] == P0}, P[t],
          {t, 0, 5}], {P0, 0, 4, 0.5}];
     Plot[Evaluate[P[t]/.solnFamily], {t, 0, 5},
          PlotRange -> {0, 4}],
     {a, 0, 4, 0.25}];
```

Do the *number* of equilibrium solutions and their *locations* confirm your earlier conjecture? (Remember that each frame includes the solution corresponding to $P_0 = 0$.)

Once we understand that the locations of the equilibrium solutions depend on the parameter a, the next step is to bypass the solution family plots altogether and simply graph the locations of the equilibrium solutions on the vertical axis with a on the horizontal axis. As Section 2.2 explains, this is called a *bifurcation diagram* for the differential equation (1'). (We'll explain the reason for this term below.) In light of your findings so far, what do you think such a diagram would look like for the equation (1')?

The following code will draw a bifurcation diagram for this equation. First we set $aP - P^2$ equal to zero and solve for P, across a range of sample values of a between 0 and 4…

```
pts = Table[Map[Point[{a, #}]&,
Select[P/.NSolve[a*P - P^2 -> 0, P], Im[#] -> 0&]],
     {a, 0, 4, 0.01}];
```

…and then plot the results:

```
Show[Graphics[pts], Axes -> True];
```

(Remember that the horizontal axis represents the parameter a, whereas the vertical axis represents the equilibrium solution(s), if any, of the equation (1') for that value of a.) Are the results what you expected? Does your bifurcation diagram show the *number* of equilibrium solutions you expected, in the *places* you expected?

Now let's add a further complication to our population model that will make the services of the computer easier to appreciate. Suppose that our population experiences harvesting at a constant rate (say by hunting or fishing), leading to the following differential equation:

$$\frac{dP}{dt} = aP - P^2 - 0.5 \qquad (1'')$$

Is it now so easy to anticipate either the *number* or the *locations* of the equilibrium solution(s)? Let's begin by making another animation similar to the one above:

```
Do[  deqn = P'[t] == a*P[t] - P[t]^2 - 0.5;
     solnFamily =
          Table[NDSolve[{deqn, P[0] == P0}, P[t],
          {t, 0, 0.9}], {P0, 0, 4, 0.2}];
     Plot[Evaluate[P[t]/.solnFamily], {t, 0, 0.9},
          PlotRange -> {0, 4}, ImageSize -> 6*72],
     {a, 0, 4, 0.1}];
```

Play the animation slowly, looking carefully for equilibrium solutions, and answer these questions:

- *How many* such solutions do you see at the *beginning* of the animation (corresponding to a near 0)?
- What about *halfway through* (corresponding to a near 2)?
- Does the *number* of equilibrium solutions seem now to vary with a?

If you answered "yes" to the third question, then you are correct! It turns out that this differential equation (1") exhibits a *bifurcation*, that is, a division, at $a = \sqrt{2} \approx 1.414$; that is, if $0 \le a < \sqrt{2}$, then the equation has no equilibrium solutions, if $a = \sqrt{2}$, then it has exactly one, and if $a > \sqrt{2}$, then it has exactly two. (Can you verify these statements using the quadratic formula?)

The value of a bifurcation diagram is that it communicates all of this information at a glance:

```
pts = Table[Map[Point[{a, #}]&,
        Select[P/.NSolve[a*P - P^2 - 0.5 -> 0, P],
            Im[#] -> 0&]], {a, 0, 4, 0.01}];
Show[Graphics[pts], Axes -> True];
```

Does the diagram confirm what you saw in the animation?

Now tweak the model (1) in ways of your own, modify the above *Mathematica* code as needed, and see what bifurcation diagrams you can draw.

Application 2.4
Implementing Euler's Method

One's understanding of a numerical algorithm is sharpened by considering its implementation in the form of a calculator or computer program. Figure 2.4.13 in the text lists TI-85 and BASIC programs implementing Euler's method to approximate the solution of the initial value problem

$$\frac{dy}{dx} = x + y, \qquad y(0) = 1 \tag{1}$$

as shown in the table of Fig 2.4.8. The comments provided in the final column of the table should make these programs intelligible even if you have little familiarity with the BASIC and Texas Instruments programming languages. To increase the number of steps (and thereby decrease the step size) you need only change the value of **N** specified in the first line of the program. To apply Euler's method to a different equation $dy/dx = f(x, y)$, you need only change the single line that calculates the function value **F**.

We illustrate below the implementation of Euler's method in systems like *Maple*, *Mathematica*, and MATLAB. For this project, you should implement Euler's method on your own calculator or in a programming language of your choice. First test your implementation by carrying through its application to the initial value problem in (1), and then apply it to solve some of the problems for Section 2.4 in the text. Then carry out the following investigation.

Famous Numbers Investigation

The problems below describe the numbers $e \approx 2.71828$, $\ln 2 \approx 0.69315$, and $\pi \approx 3.14159$ as specific values of certain initial value problem solutions. In each case, apply Euler's method with $n = 50, 100, 200, \cdots$ subintervals (doubling n each time). How many subintervals are needed to obtain — twice in succession — the correct value of the target number rounded off to 3 decimal places?

1. The number $e = y(1)$ where $y(x)$ is the solution of the initial value problem $y' = y, \quad y(0) = 1$.

2. The number $\ln 2 = y(2)$ where $y(x)$ is the solution of the initial value problem $y' = 1/x, \quad y(1) = 0$.

3. The number $\pi = y(1)$ where $y(x)$ is the solution of the initial value problem $y' = 4/(1+x^2), \quad y(0) = 0$.

Also, explain in each problem what the point is — why the approximate value of the indicated famous number is, indeed, the expected numerical result.

Using *Maple*

To apply Euler's method to the initial value problem in (1), we first define the right-hand side function $f(x, y) = x + y$ in the differential equation.

```
f := (x,y) -> x + y;
```

$$f := (x, y) \to x + y$$

To approximate the solution with initial value $y(x_0) = x_0$ on the interval $[x_0, x_f]$, we enter first the initial values of x and y and the final value of x.

```
x0 := 0:     y0 := 1:
xf := 1:
```

and then the desired number n of steps and the resulting step size h.

```
n := 10:
h := evalf((xf - x0)/n);
```

$$h := .10000$$

After we initialize the values of x and y,

```
x := x0:     y := y0:
```

Euler's method itself is implemented by the following **for**-loop, which carries out the iteration

$$y_{n+1} = y_n + h\, f(x_n, y_n), \qquad x_{n+1} = x_n + h$$

n times in successtion to take n steps across the interval from $x = x_0$ to $x = x_f$.

```
for i from 1 to n do
    k := f(x,y):           # the left-hand slope
    y := y + h*k:          # Euler step to update y
    x := x + h:            # update x
    print(x,y);            # display current values
    od:
```

.10000, 1.1000
.20000, 1.2200
.30000, 1.3620

```
.40000,   1.5282
.50000,   1.7210
.60000,   1.9431
.70000,   2.1974
.80000,   2.4871
.90000,   2.8158
1.0000,   3.1874
```

Note that x is updated after y in order that the computation $k = f(x, y)$ can use the left-hand values (with neither yet updated).

The output consists of x- and y-columns of resulting x_i- and y_i-values. In particular, we see that Euler's method with $n = 10$ steps gives $y(1) \approx 3.1874$ for the initial value problem in (1). The exact solution is $y(x) = 2e^x - x - 1$, so the actual value at $x = 1$ is $y(1) = 2e - 2 \approx 3.4366$. Thus our Euler approximation underestimates the actual value by about 7.25%.

If only the final endpoint result is wanted explicitly, then the print command can be removed from the loop and executed following it:

```
x := x0:      y := y0:        # re-initialize
for i from 1 to n do
    k := f(x,y):              # the left-hand slope
    y := y + h*k:             # Euler step to update y
    x := x + h:               # update x
    od:

print(x,y);
```

1.0000, 3.1874

For a different initial value problem, we need only enter the appropriate new function $f(x, y)$ and the desired initial and final values in the first two commands above, then re-execute the subsequent ones.

Using *Mathematica*

To apply Euler's method to the initial value problem in (1), we first define the right-hand side function $f(x, y) = x + y$ in the differential equation.

```
f[x_,y_] := x + y
```

To approximate the solution with initial value $y(x_0) = x_0$ on the interval $[x_0, x_f]$, we enter first the initial values of x and y and the final value of x.

```
x0 = 0;        y0 = 1;
xf = 1;
```

and then the desired number n of steps and the resulting step size h.

```
n = 10;
h = (xf - x0)/n  //  N

0.1
```

After we initialize the values of x and y,

```
x = x0;        y = y0;
```

Euler's method itself is implemented by the following **Do**-loop, which carries out the iteration

$$y_{n+1} = y_n + h\,f(x_n, y_n), \qquad x_{n+1} = x_n + h$$

n times in successtion to take n steps across the interval from $x = x_0$ to $x = x_f$.

```
Do[  k = f[x,y];              (* the left-hand slope      *)
     y = y + h*k;             (* Euler step to update y   *)
     x = x + h;               (* update x                 *)
     Print[x,"      ",y],     (* display updated values   *)
     {i,1,n} ]

0.1       1.1
0.2       1.22
0.3       1.362
0.4       1.5282
0.5       1.72102
0.6       1.94312
0.7       2.19743
0.8       2.48718
0.9       2.8159
1.        3.18748
```

Note that x is updated after y in order that the computation $k = f(x, y)$ can use the left-hand values (with neither yet updated).

The output consists of *x*- and *y*-columns of resulting x_i- and y_i-values. In particular, we see that Euler's method with $n = 10$ steps gives $y(1) \approx 3.1875$ for the initial value problem in (1). The exact solution is $y(x) = 2e^x - x - 1$, so the actual value at $x = 1$ is $y(1) = 2e - 2 \approx 3.4366$. Thus our Euler approximation underestimates the actual value by about 7.25%.

If only the final endpoint result is wanted explicitly, then the print command can be removed from the loop and executed following it:

```
Do[  k = f[x,y];           (* the left-hand slope      *)
     y = y + h*k;          (* Euler step to update y   *)
     x = x + h,            (* update x                 *)
     {i,1,n} ]

Print[x,"       ",y]
1.      3.18748
```

For a different initial value problem, we need only enter the appropriate new function $f(x, y)$ and the desired initial and final values in the first two commands above, then re-execute the subsequent ones.

Using MATLAB

To apply Euler's method to the initial value problem in (1), we first define the right-hand function $f(x, y)$ in the differential equation. User-defined functions in MATLAB are defined in (ASCII) text files. To define the function $f(x, y) = x + y$ we save the MATLAB function definition

```
function   yp = f(x,y)
yp = x + y;       % yp = y'
```

in the text file **f.m**.

To approximate the solution with initial value $y(x_0) = x_0$ on the interval $[x_0, x_f]$, we enter first the initial values

```
x0 = 0;    y0 = 1;
xf = 1;
```

and then the desired number n of steps and the resulting step size h.

```
n = 10;
h = (xf - x0)/n
h =
    0.1000
```

After we initialize the values of x and y,

```
x = x0;     y = y0;
```

and the column vectors **X** and **Y** of approximate values

```
X = x;     Y = y;
```

Euler's method itself is implemented by the following **for**-loop, which carries out the iteration

$$y_{n+1} = y_n + h\, f(x_n, y_n), \qquad x_{n+1} = x_n + h$$

n times in successtion to take n steps across the interval from $x = x_0$ to $x = x_f$.

```
for i = 1 : n                    % for i = 1 to n do
    k = f(x,y);                  % the left-hand slope
    y = y + h*k;                 % Euler step to update y
    x = x + h;                   % update x
    X = [X; x];                  % adjoin new x-value
    Y = [Y; y];                  % adjoin new y-value
    end
```

Note that x is updated after y in order that the computation $k = f(x, y)$ can use the left-hand values (with neither yet updated).

As output the loop above produces the resulting column vectors **X** and **Y** of x- and y-values that can be displayed simultaneously using the command

```
[X,Y]
ans =
         0    1.0000
    0.1000    1.1000
    0.2000    1.2200
    0.3000    1.3620
    0.4000    1.5282
    0.5000    1.7210
    0.6000    1.9431
    0.7000    2.1974
    0.8000    2.4872
    0.9000    2.8159
    1.0000    3.1875
```

In particular, we see that $y(1) \approx 3.1875$ for the initial value problem in (1). If only this final endpoint result is wanted explicitly, then we can simply enter

```
[X(n+1), Y(n+1)]
ans =
    1.0000    3.1875
```

The index **n+1** (instead of **n**) is required because the initial values x_0 and y_0 are the initial vector elements **X(1)** and **Y(1)**, respectively.

The exact solution of the initial value problem in (1) is $y(x) = 2e^x - x - 1$, so the actual value at $x = 1$ is $y(1) = 2e - 2 \approx 3.4366$. Thus our Euler approximation underestimates the actual value by about 7.25%.

For a different initial value problem, we need only define the appropriate function $f(x, y)$ in the file **f.m**, then enter the desired initial and final values in the first command above and re-execute the subsequent ones.

Automating Euler's Method

The **for**-loop above can be automated by saving the MATLAB function definition

```
function   [X,Y] = euler1(x,xf,y,n)
h = (xf - x)/n;                     % step size
X = x;                              % initial x
Y = y;                              % initial y
for i = 1 : n                       % begin loop
     y = y + h*f(x,y);              % Euler iteration
     x = x + h;                     % new x
     X = [X;x];                     % update x-column
     Y = [Y;y];                     % update y-column
     end                            % end loop
```

in the text file **euler1.m** (we use the name **euler1** to avoid conflict with MATLAB's built-in **euler** function). This function assumes that the function $f(x, y)$ has been defined and saved in the MATLAB file **f.m**.

The function **euler1** applies Euler's method to take n steps from x to x_f starting with the initial value y of the solution. For instance, with **f** as previously defined, the command

```
[X,Y] = euler1(0,1, 1, 10);   [X,Y]
```

is a one-liner that generates table **[X,Y]** displayed above to approximate the solution of the initial value problem $y' = x + y$, $y(0) = 1$ on the x-interval $[0, 1]$.

Application 2.5

Improved Euler Implementation

Figure 2.5.10 in the text lists TI-85 and BASIC programs implementing the improved Euler method to approximate the solution of the initial value problem

$$\frac{dy}{dx} = x + y, \quad y(0) = 1 \tag{1}$$

considered in Example 2 of Section 2.5 in the text. The comments provided in the final column should render these programs intelligible even if you have little familiarity with the BASIC and TI programming languages.

To apply the improved Euler method to a differential equation $dy/dx = f(x, y)$, one need only change the initial line of the program, in which the function f is defined. To increase the number of steps (and thereby decrease the step size) one need only change the value of N specified in the second line of the program.

We illustrate below the implementation of the improved Euler method in systems like *Maple*, *Mathematica*, and MATLAB. To begin this project, you should implement the improved Euler method on your calculator or in a programming language of your choice. Test your program by application first to the initial value problem in (1), and then to some of the problems for Section 2.5 in the text. Then carry out one or more of the following investigations.

Famous Numbers, Again

The problems below describe the numbers $e \approx 2.7182818$, $\ln 2 \approx 0.6931472$, and $\pi \approx 3.1415927$ as specific values of certain initial value problem solutions. In each case, apply the improved Euler method with $n = 10, 20, 40, \cdots$ subintervals (doubling n each time). How many subintervals are needed to obtain — twice in succession — the correct value of the target number rounded off to 5 decimal places?

1. The number $e = y(1)$ where $y(x)$ is the solution of the initial value problem $y' = y, \quad y(0) = 1$.

2. The number $\ln 2 = y(2)$ where $y(x)$ is the solution of the initial value problem $y' = 1/x, \quad y(1) = 0$.

3. The number $\pi = y(1)$ where $y(x)$ is the solution of the initial value problem $y' = 4/(1 + x^2), \quad y(0) = 0$.

Logistic Population Investigation

Apply your improved Euler program to the initial value problem

$$\frac{dy}{dt} = \frac{1}{3}y(8-y), \; y(0) = 1$$

of Example 3 in Section 2.5 in the text. In particular, verify (as claimed) that the approximate solution with step size $h=1$ levels off at $y \approx 4.3542$ rather than at the limiting value $y=8$ of the exact solution. Perhaps a table of values for $0 \le x \le 100$ will make this apparent.

For you own logistic population to investigate, you might consider the initial value problem

$$\frac{dy}{dt} = \frac{1}{n}y(m-y), \qquad y(0) = 1$$

where m and n are (for instance) the largest and smallest digits in your student ID number. Does the improved Euler approximation with step size $h = 1$ level off at the "correct" limiting value of the exact solution? If not, find a smaller value of h such that it does.

With Periodic Harvesting and Restocking

The differential equation

$$\frac{dy}{dt} = k\,y(M-y) - h\sin\left(\frac{2\pi t}{P}\right)$$

models a logistic population that is periodically harvested and restocked with period P and maximal harvesting/restocking rate h. A numerical approximation program was used to plot the typical solution curves for the case $k=M=h=P=1$ that are shown in Fig. 2.5.12 in the text. This figure suggests — though it does not suffice to prove — the existence of a threshold initial population such that

- Starting with an initial population above this threshold, the population oscillates (perhaps with period P?) about the (unharvested) stable limiting population $y(t) \equiv M$, while

- The population dies out if it starts with an initial population below this threshold.

Use an appropriate plotting utility to investigate your own logistic population with periodic harvesting and restocking (selecting typical values of the parameters k, M, h, and P). Do the observations indicated above appear to hold for your population?

Using *Maple*

To apply the improved Euler method to the initial value problem in (1), we first define the right-hand function $f(x, y) = x + y$ in the differential equation.

```
f := (x,y) -> x + y;
```

$$f := (x, y) \to x + y$$

To approximate the solution with initial value $y(x_0) = y_0$ on the interval $[x_0, x_f]$, we enter first the initial values

```
x0 := 0:      y0 := 1:      xf := 1:
```

and then the desired number n of steps and the resulting step size h.

```
n := 10:
h := evalf((xf - x0)/n);
```

$$h := .10000$$

After we initialize the values of x and y,

```
x := x0:      y := y0:
```

the improved Euler method itself is implemented by the following **`for`** loop, which carries out the iteration

$$k_1 = f(x_n, y_n), \qquad k_2 = f(x_n + h, y_n + h\,k_1),$$
$$k = \tfrac{1}{2}(k_1 + k_2),$$
$$y_{n+1} = y_n + h\,k, \qquad x_{n+1} = x_n + h$$

n times in succession to take n steps across the interval from $x = x_0$ to $x = x_f$

```
for i from 1 to n do
     k1 := f(x,y):                # the left-hand slope
     k2 := f(x+h,y+h*k1):         # the right-hand slope
     k := (k1 + k2)/2:            # the average slope
     y := y + h*k:                # Euler step to update y
     x := x + h:                  # update x
```

```
print(x,y);                    # display current values
od:
```

```
.100000,  1.11000
.200000,  1.24205
.300000,  1.39847
.400000,  1.58181
.500000,  1.79490
.600000,  2.04087
.700000,  2.32316
.800000,  2.64559
.900000,  3.01238
1.00000,  3.42818
```

Note that x is updated after y in order that the computation $k_1 = f(x, y)$ can use the left-hand values (with neither yet updated).

The output consists of x- and y-columns of resulting x_i- and y_i-values. In particular, we see that the improved Euler method with $n = 10$ steps gives $y(1) \approx 3.42818$ for the initial value problem in (1). The exact solution is $y(x) = 2e^x - x - 1$, so the actual value at $x = 1$ is $y(1) = 2e - 2 \approx 3.43656$. Thus our improved Euler approximation underestimates the actual value by about 0.24% (as compared with the 7.25% error observed in the Euler approximation of the Section 2.4 project).

If only the final endpoint result is wanted explicitly, then the print command can be removed from the loop and executed immediately following it (just as we did with the Euler loop in the Section 2.4 project). For a different initial value problem, we need only enter the appropriate new function $f(x, y)$ and the desired initial and final values in the first two commands above, then re-execute the subsequent ones.

Using *Mathematica*

To apply the improved Euler method to the initial value problem in (1), we first define the right-hand function $f(x, y) = x + y$ in the differential equation.

```
f[x_,y_] := x + y
```

To approximate the solution with initial value $y(x_0) = y_0$ on the interval $[x_0, x_f]$, we enter first the initial values

```
x0 = 0;     y0 = 1;
xf = 1;
```

and then the desired number n of steps and the resulting step size h.

```
n = 10;
h = (xf - x0)/n  //  N
0.1
```

After we initialize the values of x and y,

```
x = x0;    y = y0;
```

the improved Euler method itself is implemented by the following **Do** loop, which carries out the iteration

$$k_1 = f(x_n, y_n), \qquad k_2 = f(x_n + h, y_n + h\,k_1),$$
$$k = \tfrac{1}{2}(k_1 + k_2),$$
$$y_{n+1} = y_n + h\,k, \qquad x_{n+1} = x_n + h$$

n times in successtion to take n steps across the interval from $x = x_0$ to $x = x_f$.

```
Do[  k1 = f[x,y];            (* left-hand slope      *)
     k2 = f[x+h, y+h*k1];    (* right-hand slope     *)
      k  = (k1 + k2)/2;      (* average slope        *)
      y  = y + h*k;          (* improved Euler step *)
      x = x + h;             (* update x             *)
      Print[x,"       ",y],  (* display x and y      *)
     {i,1,n} ]

0.1     1.11
0.2     1.24205
0.3     1.39847
0.4     1.5818
0.5     1.79489
0.6     2.04086
0.7     2.32315
0.8     2.64558
0.9     3.01236
1.      3.42816
```

Note that x is updated after y in order that the computation $k_1 = f(x, y)$ use the left-hand values (with neither yet updated).

The output consists of x- and y-columns of resulting x_i- and y_i-values. In particular, we see that the improved Euler method with $n = 10$ steps gives $y(1) \approx 3.42816$ for the initial value problem in (1). The exact solution is $y(x) = 2e^x - x - 1$, so the actual value at $x = 1$ is $y(1) = 2e - 2 \approx 3.43656$. Thus our improved Euler approximation underestimates the actual value by about 0.24% (as compared with the 7.25% error observed in the Euler approximation of the Section 2.4 project).

If only the final endpoint result is wanted explicitly, then the print command can be removed from the loop and executed immediately following it (just as we did with the Euler loop in the Section 2.4 project). For a different initial value problem, we need only enter the appropriate new function $f(x, y)$ and the desired initial and final values in the first two commands above, then re-execute the subsequent ones.

Using MATLAB

To apply the improved Euler method to the initial value problem in (1), we first define the right-hand function $f(x, y)$ in the differential equation. User-defined functions in MATLAB are defined in (ASCII) text files. To define the function $f(x, y) = x + y$ we save the MATLAB function definition

```
function  yp = f(x,y)
yp = x + y;          % yp = y'
```

in the text file **f.m**.

To approximate the solution with initial value $y(x_0) = y_0$ on the interval $[x_0, x_f]$, we enter first the initial values

```
x0 = 0;       y0 = 1;       xf = 1;
```

and then the desired number n of steps and the resulting step size h.

```
n = 10;
h = (xf - x0)/n
h =
    0.1000
```

After we initialize the values of x and y,

```
x = x0;      y = y0;
```

and the column vectors **X** and **Y** of approximate values

```
X = x;       Y = y;
```

the improved Euler method itself is implemented by the following **for** loop, which carries out the iteration

$$k_1 = f(x_n, y_n), \qquad k_2 = f(x_n + h, y_n + h\,k_1),$$
$$k = \tfrac{1}{2}(k_1 + k_2),$$
$$y_{n+1} = y_n + h\,k, \qquad x_{n+1} = x_n + h$$

n times in succession to take n steps across the interval from $x = x_0$ to $x = x_f$.

```
for i = 1 : n                      % for i = 1 to n do
     k1 = f(x,y);                  % left-hand slope
     k2 = f(x+h,y+h*k1);           % right-hand slope
     k = (k1 + k2)/2;              % average slope
     y = y + h*k;                  % Euler step to update y
     x = x + h;                    % update x
     X = [X; x];                   % adjoin new x-value
     Y = [Y; y];                   % adjoin new y-value
     end
```

Note that x is updated after y in order that the computation $k = f(x, y)$ can use the left-hand values (with neither yet updated).

As output the loop above produces the resulting column vectors **X** and **Y** of x- and y-values that can be displayed simultaneously using the command

```
[X,Y]
ans =
          0    1.0000
     0.1000    1.1100
     0.2000    1.2421
     0.3000    1.3985
     0.4000    1.5818
     0.5000    1.7949
     0.6000    2.0409
     0.7000    2.3231
     0.8000    2.6456
     0.9000    3.0124
     1.0000    3.4282
```

In particular, we see that $y(1) \approx 3.4282$ for the initial value problem in (1). If only this final endpoint result is wanted explicitly, then we can simply enter

```
[X(n+1), Y(n+1)]
ans =
     1.0000    3.4282
```

The index **n+1** (instead of **n**) is required because the initial values x_0 and y_0 are the initial vector elements **X(1)** and **Y(1)**, respectively.

The exact solution of the initial value problem in (1) is $y(x) = 2e^x - x - 1$, so the actual value at $x = 1$ is $y(1) = 2e - 2 \approx 3.4366$. Thus our improved Euler approximation underestimates the actual value by about 0.24% (as compared with the 7.25% error observed in the Euler approximation of the Section 2.4 project).

For a different initial value problem, we need only define the appropriate function $f(x, y)$ in the file **f.m**, then enter the desired initial and final values in the first command above and re-execute the subsequent ones.

Automating the Improved Euler Method

The **for** loop above is a bit long for ready entry in MATLAB's command mode. The following function **impeuler** was defined by simple editing of the function **euler1** of Project 2.4.

```
function  [X,Y] = impeuler(x,xf,y,n)

h = (xf - x)/n;                  % step size
X = x;                           % initial x
Y = y;                           % initial y
for i = 1 : n                    % begin loop
    k1 = f(x,y);                 % left-hand slope
    k2 = f(x+h,y+h*k1);          % right-hand slope
    k = (k1 + k2)/2;             % average slope
    y = y + h*k;                 % improved Euler step
    x = x + h;                   % new x
    X = [X;x];                   % update x-column
    Y = [Y;y];                   % update y-column
    end                          % end loop
```

With this function saved in the text file **impeuler.m**, we need only assume also that the function $f(x, y)$ has been defined and saved in the file **f.m**.

The function **impeuler** applies the improved Euler method to take n steps from x to x_f starting with the initial value y of the solution. For instance, with **f** as previously defined, the command

```
[X,Y] = impeuler(0,1, 1, 10);
```

is a one-liner that generates the table **[X,Y]** displayed above to approximate the solution of the initial value problem $y' = x + y$, $y(0) = 1$ on the x-interval [0, 1].

Application 2.6

Runge-Kutta Implementation

Figure 2.6.11 in the text lists TI-85 and BASIC programs implementing the Runge-Kutta method to approximate the solution of the initial value problem

$$\frac{dy}{dx} = x + y, \quad y(0) = 1 \tag{1}$$

considered in Example 1 of Section 2.6. Even if you have little familiarity with the BASIC and TI programming languages, the comments provided in the final column should make it clear how each is implementing the Runge-Kutta iteration

$$\begin{aligned} k_1 &= f(x_n, y_n), \\ k_2 &= f(x_n + \tfrac{1}{2}h, y_n + \tfrac{1}{2}h\, k_1), \\ k_3 &= f(x_n + \tfrac{1}{2}h, y_n + \tfrac{1}{2}h\, k_2) \\ k_4 &= f(x_n + h, y_n + h\, k_3) \\ k &= \tfrac{1}{6}\left(k_1 + 2k_2 + 2k_2 + k_4\right), \\ y_{n+1} &= y_n + h\, k, \\ x_{n+1} &= x_n + h \end{aligned} \tag{2}$$

to make the step from (x_n, y_n) to (x_{n+1}, y_{n+1}).

To apply the Runge-Kutta method to a differential equation $dy/dx = f(x, y)$, one need only change the initial line of the program, in which the function f is defined. To increase the number of steps (and thereby decrease the step size) one need only change the value of N specified in the second line of the program.

We illustrate below the implementation of the Runge-Kutta method in systems like *Maple*, *Mathematica*, and MATLAB. To begin this project, you should implement the Runge-Kutta method on your calculator or in a programming language of your choice. First test your program by carrying through its application to the initial value problem in (1), and then apply it to solve some of the problems for Section 2.6 in the text. Then carry out the following "famous numbers" and "skydiver's descent" investigations.

A. Famous Numbers, One Last Time

The problems below describe the numbers $e \approx 2.71828182846$, $\ln 2 \approx 0.69314718056$, and $\pi \approx 3.14159265359$ as specific values of certain initial value problem solutions. In each case, apply the improved Euler method with $n = 10, 20, 40, \cdots$ subintervals (doubling n each time). How many subintervals are needed to obtain — twice in succession — the correct value of the target number rounded off accurate to 9 decimal places?

1. The number $e = y(1)$ where $y(x)$ is the solution of the initial value problem $y' = y, \quad y(0) = 1$.

2. The number $\ln 2 = y(2)$ where $y(x)$ is the solution of the initial value problem $y' = 1/x, \quad y(1) = 0$.

3. The number $\pi = y(1)$ where $y(x)$ is the solution of the initial value problem $y' = 4/(1+x^2), \quad y(0) = 0$.

B. The Skydiver's Descent

Recall the 60-kg skydiver of Example 3 in Section 2.6 of the text. She falls vertically with initial velocity zero from an initial height of 5 kilometers, and experiences an upward force F_R of air resistance given in terms of her velocity v (in m/s) by $F_R = 0.0096\left(100v + 10v^2 + v^3\right)$. Then her downward velocity function $v(t) > 0$ satisfies the $ma = F$ equation

$$60\frac{dv}{dt} = 60 \times 9.8 - 0.0096(100v + 10v^2 + v^3) \tag{3}$$

(with $m = 60$ and $g = 9.8$ in mks units), and hence the initial value problem

$$\frac{dv}{dt} = 9.8 - 0.00016(100v + 10v^2 + v^3), \qquad v(0) = 0. \tag{4}$$

The skydiver reaches her terminal velocity when the forces of gravity and air resistance balance, so $dv/dt = 0$. We saw in Example 3 that this terminal velocity is approximately 35.578 m/s.

First apply the Runge-Kutta method with step size $h = 0.1$ to generate the table of velocities shown in Fig. 2.6.8 of the text, where we see that the skydiver has achieved her terminal velocity after 20 seconds of free fall. Then approximate the skydiver's successive (downward) positions $y_1, y_2, y_3, \ldots$ by beginning with the initial position $y(0) = y_0 = 0$ and calculating

$$y_{n+1} = y_n + v_n h + \tfrac{1}{2}a_n h^2 \tag{5}$$

$(n = 1, 2, 3, \ldots)$ where $a_n = v'(t_n)$ is the particle's approximate acceleration at time t_n.

Formula (5) would give the correct increment (from y_n to y_{n+1}) *if* the acceleration a_n remained constant during the time interval $[t_n, t_{n+1}]$. You should obtain the position data shown in Fig. 2.6.13 in the text, where it appears that the skydiver falls 629.866 meters during her first 20 seconds of descent. She then free falls the remaining 4370.134 meters to the ground at her terminal speed of 35.578 meters per second. Hence her total time of descent is $20 + 4370.134/35.578 = 142.833$ seconds, or about 2 min 23 sec.

For an individual problem to solve using whatever computational system is available to you, analyze your own skydive (perhaps from a different height), using your own mass m (in kilograms) and a plausible air resistance force of the form $F_R = av + bv^2 + cv^3$. For instance, for coefficients of the same magnitude used above, you could take $a = 0.01p^2$, $b = 0.01q$, and $c = 0.01$ where p and q are the two largest digits in your student ID number.

Using *Maple*

To apply the Runge-Kutta method to the initial value problem in (1), we first define as usual the right-hand function in the differential equation.

```
f := (x,y) -> x + y:
```

To approximate the solution with initial value $y(x_0) = y_0$ on the interval $[x_0, x_f]$, we enter first the initial values

```
x0 := 0:     y0 := 1:
xf := 1:
```

and then the desired number n of steps and the resulting step size h.

```
n := 10:
h := evalf((xf - x0)/n):
```

After we initialize the values of x and y,

```
x := x0:     y := y0:
```

the Runge-Kutta method itself is implemented by the **for** loop below, which carries out the iteration in (2) n times in succession to take n steps across the interval from $x = x_0$ to $x = x_f$

```
for i from 1 to n do
   k1 := f(x,y):              # the left-hand slope
   k2 := f(x+h/2,y+h*k1/2):   # 1st midpoint slope
   k3 := f(x+h/2,y+h*k2/2):   # 2nd midpoint slope
```

```
k4 := f(x+h,y+h*k3):          # the right-hand slope
k  := (k1+2*k2+2*k3+k4)/6:    # the average slope
y  := y + h*k:                # R-K step to update y
x  := x + h:                  # update x
print(x,y);                   # display current values
od:
```

```
             .1000000,    1.110342
             .2000000,    1.242806
             .3000000,    1.399718
             .4000000,    1.583650
             .5000000,    1.797443
             .6000000,    2.044238
             .7000000,    2.327506
             .8000000,    2.651082
             .9000000,    3.019206
             1.000000,    3.436563
```

Thus the Runge-Kutta method with $n = 10$ steps gives $y(1) \approx 3.436563$ for the initial value problem in (1). The actual value of the exact solution $y(x) = 2e^x - x - 1$, at $x = 1$ is $y(1) = 2e - 2 \approx 3.436564$, so with only 10 steps the Runge-Kutta gives nearly 6-decimal place accuracy!

Automating the Runge-Kutta Method

The following *Maple* procedure makes n steps from x_0 to x_f to approximate a solution of the initial value problem $y' = f(x,y)$, $y(x_0) = y_0$.

```
rk := proc(x0,xf,y0,n)
      local x,y,h,i,k1,k2,k3,k4,k,R;
      x := x0;
      y := y0;
      R := x,y;
      h := evalf((xf-x0)/n);
      for i from 1 by 1 to n do
         k1 := f(x,y);
           k2 := f(x+h/2,y+h*k1/2);
           k3 := f(x+h/2,y+h*k2/2);
           k4 := f(x+h,y+h*k3);
          k := (k1+2*k2+2*k3+k4)/6;
          y := y + h*k;
          x := x + h;
          R := R,x,y
          od
      end:
```

The output of **rk** is the list **R** of successive values $x_0, y_0, x_1, y_1, x_2, y_2, \ldots, x_n, y_n$ which we can easily print as a table:

```
R := rk(0,1,1,10):
     for i from 1 by 1 to 11 do
     print(R[2*i-1],R[2*i])   od;
```

```
                 0,      1
       .1000000000      1.110341667
       .2000000000,     1.242805142
       .3000000000,     1.399716994
       .4000000000,     1.583648480
       .5000000000,     1.797441277
       .6000000000,     2.044235924
       .7000000000,     2.327503253
       .8000000000,     2.651079126
       .9000000000,     3.019202827
       1.000000000,     3.436559488
```

The Skydiver Problem

We now redefine f as the skydiver's acceleration function.

```
f := (t,v) -> 9.8 - 0.00016*(100*v + 10*v^2 + v^3):
```

The following commands take $k = 200$ steps from $t = 0$ to $t = 20$, and assemble the lists **T** of successive times and **V** of successive velocities.

```
k := 200:
R := rk(0,20,0,k):
t := 0:   v := 0:     # initial time and velocity
T := t:   V := v:     # initialize lists
for i from 1 by 1 to k do
T := T,R[2*i+1];
V := V,R[2*i+2]
od:
```

The print loop

```
for i from 1 by 10 to 201 do
     print(T[i],V[i])   od;
```

then displays the time-velocity data shown in Fig. 2.6.8 in the text. Finally we use (5) to calculate the corresponding list **Y** of downward positions of the skydiver.

```
Y[1] := 0:   # initial position
h := 0.1:    # step size
```

```
for n from 1 by 1 to k do
     a := f(T[n],V[n]);
     Y[n+1]  := Y[n] + V[n]*h + 0.5*a*h^2
     od:
```

Then the print loop

```
for i from 1 by 20 to 201 do
     print(T[i],V[i],Y[i])  od;
```

displays the time-velocity-position data shown in Fig. 2.6.13 in the text. We see that the skydiver falls 629.866 meters during her first 20 seconds of descent.

Using *Mathematica*

To apply the Runge-Kutta method to the initial value problem in (1), we first define as usual the right-hand function in the differential equation.

```
f[x_,y_]  := x + y
```

To approximate the solution with initial value $y(x_0) = y_0$ on the interval $[x_0, x_f]$, we enter first the initial values

```
x0 = 0;       y0 = 1;
xf = 1;
```

and then the desired number n of steps and the resulting step size h.

```
n = 10;
h = (xf - x0)/n  //  N;
```

After we initialize the values of x and y,

```
x = x0;    y = y0;
```

the Runge-Kutta method itself is implemented by the following `Do` loop, which carries out the iteration in (2) n times in succession to take n steps across the interval from $x = x_0$ to $x = x_f$

```
Do[  k1 = f[x,y];                (* left-hand slope   *)
     k2 = f[x+h/2, y+h*k1/2];    (* 1st midpt slope   *)
     k3 = f[x+h/2, y+h*k2/2];    (* 2nd midpt slope   *)
     k4 = f[x + h, y + h*k3];    (* right-hand slope *)
     k  = (k1+2k2+2k3+k4)/6;     (* average slope     *)
     y = y + h*k;                (* Runge-Kutta step *)
     x = x + h;                  (* update x          *)
```

```
Print[x,"        ",y],          (* display x and y  *)
{i,1,n} ]
```

```
0.1     1.11034
0.2     1.24281
0.3     1.39972
0.4     1.58365
0.5     1.79744
0.6     2.04424
0.7     2.3275
0.8     2.65108
0.9     3.0192
1.      3.43656
```

Thus the Runge-Kutta method with $n = 10$ steps gives $y(1) \approx 3.43656$ for the initial value problem in (1). The actual value of the exact solution $y(x) = 2e^x - x - 1$, at $x = 1$ is $y(1) = 2e - 2 \approx 3.436564$, so with only 10 steps the Runge-Kutta gives 5-decimal place accuracy!

Automating the Runge-Kutta Method

The following *Mathematica* function makes n steps from x_0 to x_f to approximate a solution of the initial value problem $y' = f(x,y)$, $y(x_0) = y_0$.

```
rk[x0_, xf_, y0_, n_] :=
     Module[{h,x,y,X,Y,i,k1,k2,k3,k4,k},
     h = (xf - x0)/n;
     x = x0; y = y0;
     X = {x};  Y = {y};
     Do[k1 = f[x, y];
          k2 = f[x+h/2, y+h*k1/2];
          k3 = f[x+h/2, y+h*k2/2];
          k4 = f[x+h, y+h*k3];
          k = (k1+2*k2+2*k3 k4)/6;
          y = y + h*k;
          x = x + h;
     X = Append[X, x]; Y = Append[Y, y],
     {i, 1, n}];
     {X, Y}]
```

The output of **rk** is the pair **{X,Y}** of lists **X** and **Y** of successive x- and y-values, which we can easily print in table format:

```
{X, Y} = rk[0., 1., 1, 10];
TableForm[Transpose[{X, Y}]]
```

```
0.      1
0.1     1.11034
0.2     1.24281
0.3     1.39972
0.4     1.58365
0.5     1.79744
0.6     2.04424
0.7     2.3275
0.8     2.65108
0.9     3.0192
1.0     3.43656
```

The Skydiver Problem

We now redefine f as the skydiver's acceleration function.

```
f[t_,v_] := 9.8 - 0.00016 (100v + 10v^2 + v^3)
```

We can use **rk** to take $k = 200$ steps from $t = 0$ to $t = 20$ and calculate the lists **T** of successive times and **V** of successive velocities of the skydiver.

```
k = 200;
{T, V} = rk[0., 20, 0, k];
```

Then

```
{T[[k+1]], V[[k+1]]}
{20., 35.5779}
```

verifies that the terminal velocity of 35.578 m/s has been achieved after 20 seconds. Finally we use (5) to calculate the corresponding list **Y** of downward positions.

```
Y = Table[0, {i, 1, k + 1}];
h = 0.1;
Do[  a = f[T[[n]], V[[n]]];
     Y[[n + 1]] = Y[[n]] + V[[n]]*h + 0.5*a*h^2,
     {n, 1, k}];
```

Then the result

```
{T[[k+1]], V[[k+1]], Y[[k+1]]}
{20., 35.5779, 629.866}
```

indicates that the skydiver falls 629.866 meters during her first 20 seconds of descent.

Using MATLAB

To apply the Runge-Kutta method to the initial value problem in (1), suppose as usual that the right-hand side function

```
function   yp = f(x,y)
yp = x + y;     % yp = y'
```

in the differential equation has been defined and saved in the text file **f.m**.

To approximate the solution with initial value $y(x_0) = y_0$ on the interval $[x_0, x_f]$, we enter first the initial values

```
x0 = 0;       y0 = 1;       xf = 1;
```

and then the desired number n of steps and the resulting step size h.

```
n = 10;
h = (xf - x0)/n;
```

After we initialize the values of x and y,

```
x = x0;      y = y0;
```

and the column vectors **X** and **Y** of approximate values

```
X = x;       Y = y;
```

the Runge-Kutta method itself is implemented by the following **for** loop, which carries out the iteration (2) n times in succession to take n steps across the interval from $x = x_0$ to $x = x_f$.

```
for i = 1 : n                        % for i = 1 to n do
    k1 = f(x,y);                     % left-hand slope
    k2 = f(x+h/2,y+h*k1/2);          % 1st midpoint slope
    k3 = f(x+h/2,y+h*k2/2);          % 2nd midpoint slope
    k4 = f(x+h,y+h*k3):              % the right-hand slope
     k = (k1+2*k2+2*k3+k4)/6;        % the average slope
     y = y + h*k;                    % R-K step to update y
     x = x + h;                      % update x
     X = [X; x];                     % adjoin new x-value
     Y = [Y; y];                     % adjoin new y-value
     end
```

Note that x is updated after y in order that the computation $k = f(x, y)$ can use the left-hand values (with neither yet updated).

As output, the loop above produces the resulting column vectors **X** and **Y** of x- and y-values that can be displayed simultaneously using the command

```
format long
[X,Y]
```

```
ans =
                   0   1.00000000000000
    0.10000000000000   1.11034166666667
    0.20000000000000   1.24280514170139
    0.30000000000000   1.39971699412508
    0.40000000000000   1.58364848016137
    0.50000000000000   1.79744127719368
    0.60000000000000   2.04423592418387
    0.70000000000000   2.32750325319355
    0.80000000000000   2.65107912658463
    0.90000000000000   3.01920282756014
    1.00000000000000   3.43655948827033
```

which displays the column vectors **X** and **Y** side-by-side. Thus the Runge-Kutta method with $n = 10$ steps gives $y(1) \approx 3.43656$ for the initial value problem in (1). The actual value of the exact solution $y(x) = 2e^x - x - 1$ at $x = 1$ is $y(1) = 2e - 2 \approx 3.436564$, so with only 10 steps the Runge-Kutta gives 5-decimal place accuracy!

For a different initial value problem, we need only define the appropriate function $f(x, y)$ in the file **f.m**, then enter the desired initial and final values in the first command above and re-execute the subsequent ones.

Automating the Runge-Kutta Method

The following MATLAB function makes n steps from x_0 to x_f to approximate a solution of the initial value problem $y' = f(x,y), \quad y(x_0) = y_0$.

```
function   [X,Y] = rk(x,xf,y,n)

h = (xf - x)/n;                 % step size
X = x;                          % initial x
Y = y;                          % initial y
for i = 1 : n                   % begin loop
   k1 = f(x,y);                 % left-hand slope
   k2 = f(x+h/2,y+h*k1/2);      % 1st midpoint slope
   k3 = f(x+h/2,y+h*k2/2);      % 2nd midpoint slope
   k4 = f(x+h,y+h*k3);          % right-hand slope
   k = (k1+2*k2+2*k3+k4)/6;     % average slope
   y = y + h*k;                 % Runge-Kutta step
   x = x + h;                   % new x
```

```
X = [X;x];                          % update x-column
Y = [Y;y];                          % update y-column
end                                 % end loop
```

With this function saved in the text file **rk.m**, we need only assume also that the function $f(x, y)$ has been defined and saved in the file **f.m**.

The function **rk** applies the Runge-Kutta method to take n steps from x to x_f starting with the initial value y of the solution. The output of **rk** consists of the pairs **X** and **Y** of column vectors of successive x- and y-values. For instance, with **f** as previously defined, the command

```
[X,Y] = rk(0, 1, 1, 10);   [X,Y]
```

is a one-liner that generates the table **[X,Y]** displayed above to approximate the solution of the initial value problem $y' = x + y$, $y(0) = 1$ on the x-interval $[0, 1]$.

The Skydiver's Descent

To apply the Runge-Kutta method to the initial value problem in (3), we re-define the MATLAB function **f.m** as the skydiver's acceleration function:

```
function  vp = f(t,v)
vp = 9.8 - 0.00016*(100*v + 10*v^2 + v^3);
```

Then the commands

```
k = 200;                        % 200 subintervals
[T,V] = rk(0,20,0,k);           % Runge-Kutta approximation
```

take $k = 200$ steps from $t = 0$ to $t = 20$ and calculate the column vectors **T** of successive times and **V** of successive velocities. Hence the command

```
[t(1:10:k+1);v(1:10:k+1)]   % Display every 10th entry
```

produces the table time-velocity data shown in Fig. 2.6.8 of the text, and verifies that the skydiver has achieved her terminal velocity 35.578 m/s after 20 seconds. Finally, the commands

```
Y = zeros(k+1,1);                             % initialize Y
h = 0.1;                                      % step size
for n = 1:k                                   % for n=1 to k
   a = f(T(n),V(n));                          % acceleration
   Y(n+1) = Y(n) + V(n)*h + 0.5*a*h^2;        % Equation (5)
   end                                        % end loop
```

```
[T(1:20:k+1),V(1:20:k+1),Y(1:20:k+1)]   % each 20th entry
```

calculate the corresponding column vector **Y** successive downward positions of the skydiver, and then display the time-velocity-position data shown in the table of Fig. 2.6.13 in the text. We see that the skydiver falls 629.866 meters during her first 20 seconds of descent.

Chapter 3

Linear Equations of Higher Order

Application 3.1
Plotting Second-Order Solution Families

This application deals with the computer-plotting of solution families like those illustrated in Figs. 3.1.6 and Fig. 3.1.7 in the text. Show first that the general solution of the differential equation

$$y'' + 3y' + 2y = 0 \tag{1}$$

is

$$y(x) = c_1 e^{-x} + c_2 e^{-2x}. \tag{2}$$

Then show that the particular solution of Eq. (1) satisfying the initial conditions $y(0) = a,\ y'(0) = b$ is

$$y(x) = (2a + b)e^{-x} - (a + b)e^{-2x} \tag{3}$$

- For Fig. 3.1.6, substitution of $a = 1$ in (3) gives

$$y(x) = (b + 2)e^{-x} - (b + 1)e^{-2x}. \tag{4}$$

for the solution curve through the point $(0,1)$ with initial slope $y'(0) = b$.

- For Fig. 3.1.7, substitution of $b = 1$ in (3) gives

$$y(x) = (2a + 1)e^{-x} - (a + 1)e^{-2x}. \tag{5}$$

for the solution curve through the point $(0,a)$ with initial slope $y'(0) = 1$.

In the technology-specific sections following the problems below, we illustrate the use of computer systems like *Maple*, *Mathematica*, and MATLAB to plot simultaneously a family of solution curves like those defined by (4) or (5). Start by reproducing Figs. 3.1.6 and 3.1.7 in the text. Then, for each of the following differential equations, construct both a family of different solution curves satisfying $y(0)=1$ and a family of different solution curves satisfying the initial condition $y'(0)=1$.

1. $y''-y=0$

2. $y''-3y'+2=0$

3. $2y''+3y'+y=0$

4. $y''+y=0$ (with general solution $y(x)=c_1\cos x+c_2\sin x$)

5. $y''+2y'+2y=0$ (with general solution $y(x)=e^{-x}(c_1\cos x+c_2\sin x)$)

Using *Maple*

Using Eq. (4), the particular solution with $y(0)=1,\ y'(0)=b$ is defined by

```
partSoln := (b+2)*exp(-x)-(b+1)*exp(-2*x);
```

$$partSoln := (b+2)e^{(-x)}-(b+1)e^{-(2x)}$$

The set of such particular solutions with initial slopes $b=-5,-4,-3,\ldots,4,\ 5$ is then defined by

```
curves := {seq(partSoln, b =-5..5)}:
```

We plot these 11 curves simultaneously on the *x*-interval (–1, 5) with the single command

```
plot(curves, x =-1..5, y =-1..3);
```

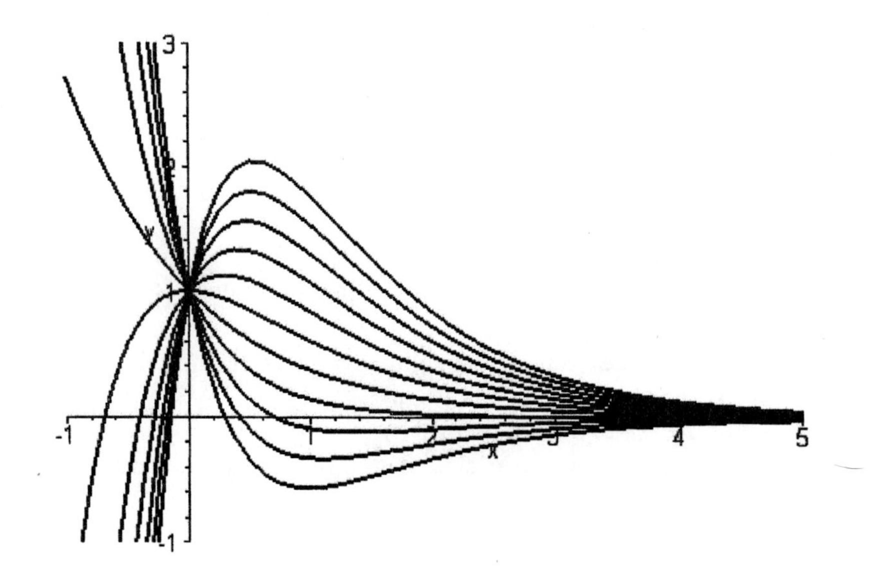

Using *Mathematica*

Using Eq. (5), the particular solution with $y(0) = a,\ y'(0) = 1$ is defined by

```
partSoln = (2a + 1) Exp[-x] - (a + 1) Exp[-2x]
```

$(2a+1)e^{-x} - (a+1)e^{-2x}$

The set of such particular solutions with initial slopes $a = -5, -4, -3, \ldots, 4, 5$ is then defined by

```
curves = Table[ partSoln, {a,-5,5} ];
```

We plot these 11 curves simultaneously on the x-interval $(-1, 5)$ with the single command

```
Plot[Evaluate[curves], {x,-2,4}, PlotRange->{-6,6}]
```

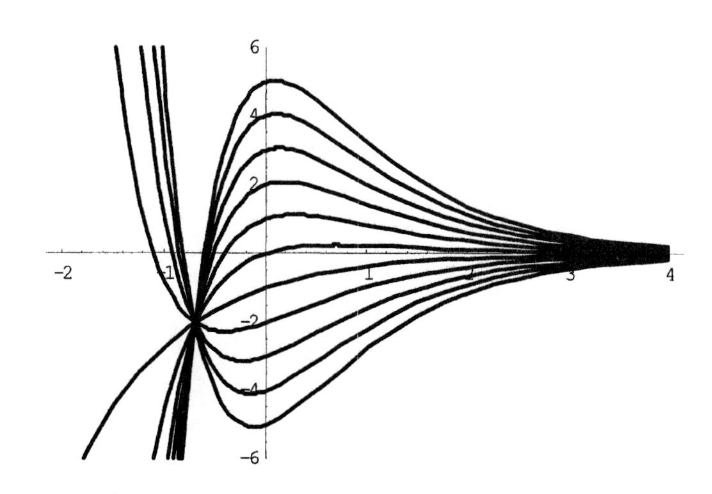

Using MATLAB

Using Eq. (5), the particular solution with $y(0) = y'(0) = a$ is defined by

$$y(x) = 3a\,e^{-x} - 2a\,e^{-2x}.$$

We can plot the 11 solution curves with $a = -5, -4, -3, \ldots, 4, 5$ on the interval

```
x = -1 : 0.02 : 5;   % x-vector from x=-1 to x=5
```

with the single **for** loop

```
for a = -5 : 5          % for a = -1 to 1 do
    y = 3*a*exp(-x) - 2*a*exp(-2*x);
    plot(x,y,'k')
    axis([-1 5 -7 7]), hold on
    end
```

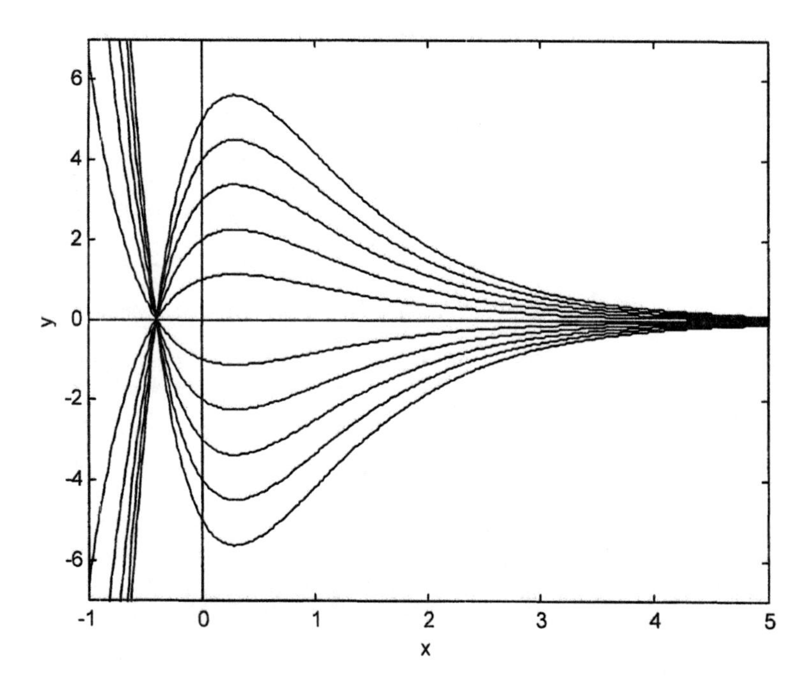

Further Investigation

Have you noticed anything common to all three of the plots of solution families shown above (as well as several of the plots you made for differential equations **1-5** above)? In this section we will examine this phenomenon more closely—and try to identify the circumstances under which it occurs.

First, review the plots you made for differential equations **1-5** for the case where $y'(0)$ is held fixed at 1, and answer the following question:

> For which of these solution families does it seem that *all the solution curves meet at a common point in the plane*?

(Be prepared to change the "viewing rectangle" if you think something significant might be going on off-screen.)

To see what makes this phenomenon occur with one differential equation and not another, list the *characteristic roots* for each of the equations **1-5** listed at the beginning. Can you make a conjecture about when the phenomenon occurs?

To test your conjecture (or to help you form one), plot solution families like the ones above for the following differential equations (of course with the same initial conditions $y'(0)=1$, $y(0)=a$, and with a ranging from -5 to 5):

- $y''+y'-2y=0$
- $y''-y=0$
- $y''-y'=0$

By now you are probably convinced that there is a theorem in here somewhere, and indeed there is! Can you prove that (as one example) the phenomenon we have observed always occurs when the characteristic roots are *real, distinct, and of the same sign*? As a bonus, your proof should also predict for you—in terms of the characteristic roots—the *value* of x at which the solution curves meet; compare this prediction with the graphs you found above. (*Hint*: Call the roots r_1 and r_2 and write the solution y of the initial value problem explicitly in terms of $a=y(0)$. There is a point at which the solution curves meet if and only if there is an x at which y does not depend upon a, that is, at which $\frac{\partial y}{\partial a}=0$.)

Application 3.2
Plotting Third-Order Solution Families

This application deals with the computer-plotting of solution families like those illustrated in Figs. 3.2.2 through 3.2.4 in the text. We know from Example 6 in Section 3.2 that the general solution of

$$y^{(3)} + 3y'' + 4y' + 12y = 0 \tag{1}$$

is

$$y(x) = c_1 e^{-3x} + c_2 \cos 2x + c_3 \sin 2x . \tag{2}$$

Then use the method of Example 6 (or a computer algebra system) to show that the particular solution of (1) satisfying the initial conditions $y(0) = a,\ y'(0) = b$ and $y''(0) = c$ is

$$y(x) = \frac{1}{26}\left[(8a+2c)e^{-3x} + (18a-2c)\cos 2x + (12a+13b+3c)\sin 2x\right]. \tag{3}$$

- For Fig. 3.2.2, substitution of $b = c = 0$ in (3) gives

$$y(x) = \frac{a}{13}\left(4e^{-3x} + 9\cos 2x + 6\sin 2x\right). \tag{4}$$

- For Fig. 3.2.3, substitution of $a = c = 0$ in (3) gives

$$y(x) = \frac{b}{2}\sin 2x . \tag{5}$$

- For Fig. 3.2.4, substitution of $a = b = 0$ in (3) gives

$$y(x) = \frac{c}{26}\left(2e^{-3x} - 2\cos 2x + 3\sin 2x\right). \tag{6}$$

In the following sections, we illustrate the use of computer systems like *Maple*, *Mathematica*, and MATLAB to plot simultaneously a family of solution curves like those defined by (4), (5), or (6). Start by reproducing Figs. 3.2.2 –3.2.4. Then plot similar families of solution curves for the differential equations in Problems 13–20 of Section 3.2 in the text.

Using *Maple*

Using Eq. (4), the particular solution of (1) with $y(0) = a,\ y'(0) = 0$ and $y''(0) = 0$ is defined by

```
partSoln := a*(4*exp(-3*x)+9*cos(2*x)+6*sin(2*x))/13;
```

$$partSoln := \frac{1}{13}a\left(4\,e^{(-3x)} + 9\cos(2x) + 6\sin(2x)\right)$$

The set of such particular solutions with initial slopes $b = -3, -2, \ldots, 2, 3$ is then defined by

```
curves := {seq(partSoln, b =-3..3)}:
```

We plot these 7 curves simultaneously on the x-interval $(-1.5, 5)$ with the single command

```
plot(curves, x =-1.5..5, y =-5..5);
```

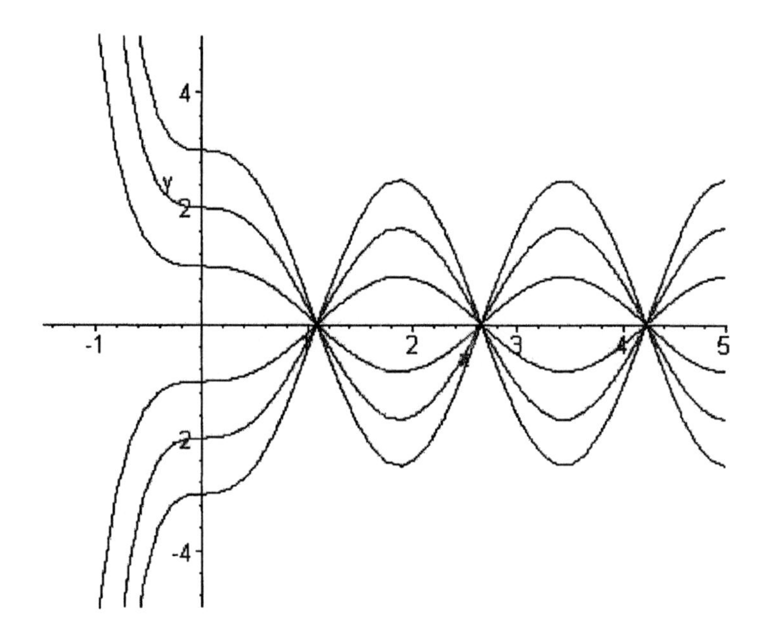

Using *Mathematica*

Using Eq. (6), the particular solution of (1) with $y(0) = 0,\ y'(0) = 0$ and $y''(0) = c$ is defined by

```
partSoln = c(2 Exp[-3x] - 2 Cos[2x] + 3 Sin[2x])/26
```

$$\frac{1}{26}c\left(2e^{-3x}-2\cos(2x)+3\sin(2x)\right)$$

The set of such particular solutions with initial slopes $c = -3, -2, \ldots, 2, 3$ is then defined by

```
curves = Table[partSoln, {c, -3, 3}];
```

We plot these 7 curves simultaneously on the x-interval $(-1.5, 5)$ with the single command

```
Plot[Evaluate[curves], {x,-1.5,5}, PlotRange->{-1,1}];
```

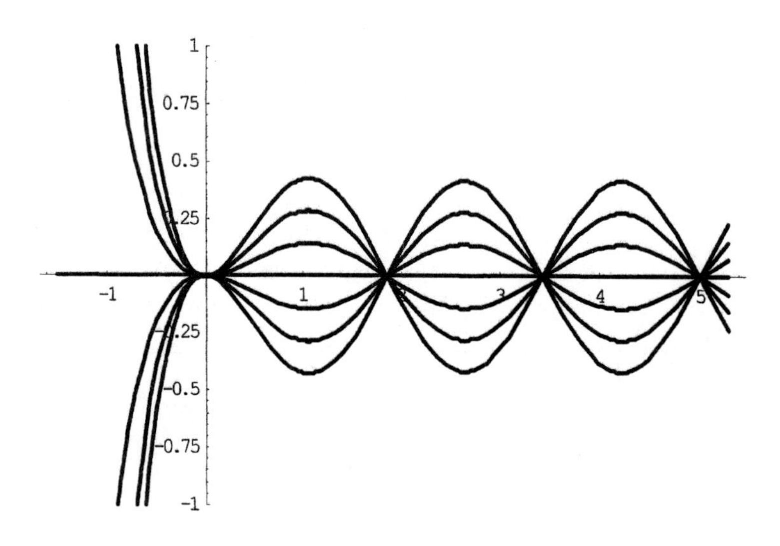

Using MATLAB

Using Eq. (3), the particular solution with $y(0) = y'(0) = y''(0) = a$ is defined by

$$y(x) = \frac{a}{13}\left(5e^{-3x} + 8\cos 2x + 14\sin 2x\right).$$

We can plot the 7 solution curves with $a = -6, -4, \ldots, 4, 6$ on the interval

```
x = -1.5 : 0.02 : 5;    % x-vector from x=-1.5 to x=5
```

with the single **for** loop

```
for a = -6 : 2 : 6
    c1 = 5*a/13;   c2 = 8*a/13;   c3 = 14*a/13;
    y = c1*exp(-3*x) + c2*cos(2*x) + c3*sin(2*x);
    plot(x,y,'k')
    axis([-1.5 5 -10 10]), hold on
    end
```

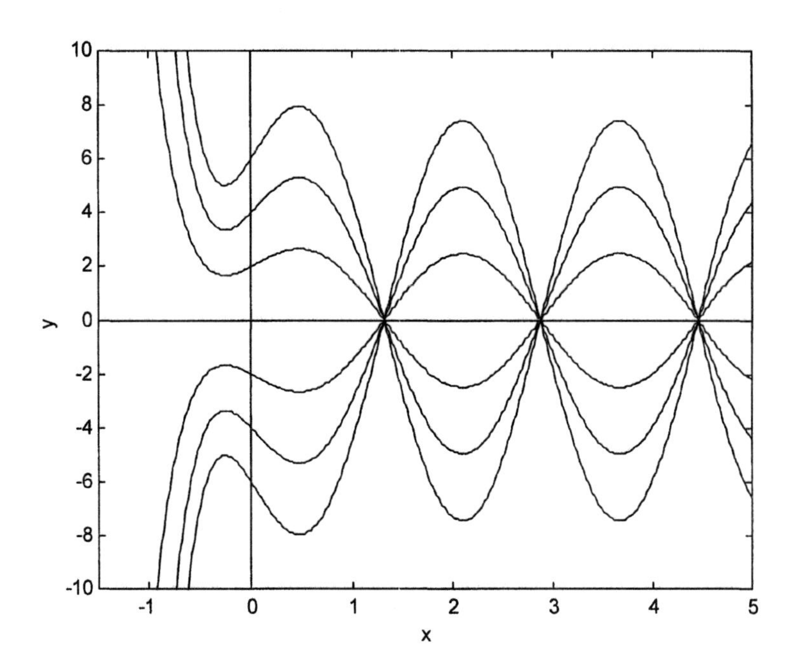

Application 3.3
Approximate Solutions of Linear Equations

Section 3.3 in the text shows that the problem of solving a homogeneous linear differential equation reduces to solving its characteristic (polynomial) equation. For this and similar purposes, polynomial-solving utilities are now a common feature of calculator and computer systems, and can be used to solve a characteristic equation numerically even when no simple and explicit factorization is evident or even possible. For instance, suppose that we want to solve the homogeneous linear differential equation

$$y''' - 3\,y'' + y = 0 \tag{1}$$

with characteristic equation

$$r^3 - 3r^2 + 1 = 0. \tag{2}$$

A computer algebra system provides the three solutions of this cubic equation in the form

$$r_1 = 1 + u + \frac{1}{u}, \quad r_2 = 1 - u^2 - uv^3, \quad r_3 = 1 - u^4 - \frac{v^3}{u}$$

where

$$u = \sqrt[3]{\frac{1 + i\sqrt{3}}{2}} \quad \text{and} \quad v = \sqrt[3]{\frac{1 - i\sqrt{3}}{2}}$$

involving cube roots of complex numbers. For instance,

$$r_1 = 1 + \sqrt[3]{\frac{1 + i\sqrt{3}}{2}} + \sqrt[3]{\frac{2}{1 + i\sqrt{3}}}.$$

One could write a general solution of (1) in the form

$$y(x) = c_1 e^{r_1 x} + c_2 e^{r_2 x} + c_3 e^{r_3 x}. \tag{3}$$

However, this general solution would seem needlessly complex, inasmuch as the graph of the function $f(x) = x^3 - 3x^2 + 1$ shows that all three roots of (2) actually are real numbers!

Indeed, with a calculator (like the TI-89) having a built-in polynomial solver, we can simply enter the coefficients $1, -3, 0, 1$ of this cubic polynomial and get the three (approximate) roots $r = -0.5321, 0.6527, 2.8794$ at the press of a key. Computer systems like *Maple*, *Mathematica*, and MATLAB also have built-in polynomial solvers that can provide such numerical solutions.

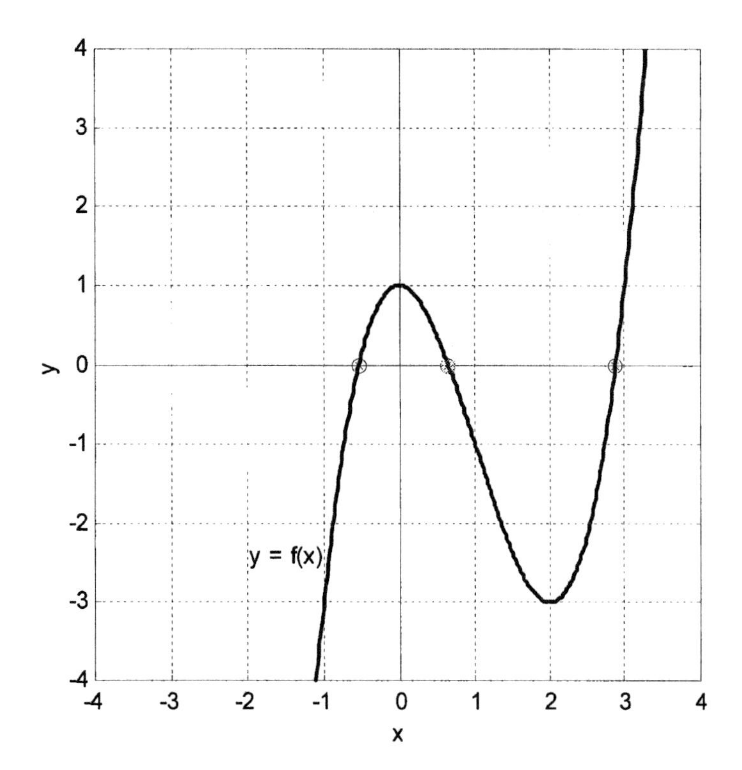

However we find these numerical roots of (2), it follows by substitution in (3) that a general solution of the differential equation in (1) is given (approximately) by

$$y(x) \;=\; c_1 e^{-0.5321x} + c_2 e^{0.6527x} + c_3 e^{2.8794x}. \tag{4}$$

Computer algebra systems also offer simple DE solver commands for the explicit solution of differential equations. It is interesting to compare the symbolic solutions produced by such DE solvers with explicit numerical solutions of the form in (4).

Use calculator or computer methods to find general solutions (in approximate numerical form) of the following differential equations. Compare the results obtained using a polynomial solver and using a DE solver.

1. $y''' - 3y' + y = 0$

2. $y''' + 3y'' - 3y = 0$

3. $y''' + y' + y = 0$

4. $y''' + 3y' + 5y = 0$

5. $y^{(4)} + 2y''' - 3y = 0$

6. $y^{(4)} + 3y' - 4y = 0$

Using *Maple*

The characteristic equation in (2) is defined by

```
charEq := r^3 - 3*r^2 + 1 = 0:
```

The command

```
soln := solve(charEq, r);
```

yields complex "Cardan formula expressions" like those exhibited previously, but the floating point evaluation

```
soln := [evalf(soln,5)];
```

$$soln := [2.8794{+}.00001\,I,\ -.53209{+}.00002\,I,\ .65271-.00002\,I]$$

after deletion of the imaginary round-off errors,

```
soln := map(Re,soln);
```

$$soln := [2.8794, -.53209, .65271]$$

gives the three approximate characteristic roots mentioned before. We can now assemble the approximate solution

```
y = sum(c[i]*exp(soln[i]*x),i=1..3);
```

$$y = c_1\,e^{(2.8794\,x)} + c_2\,e^{(-.53209\,x)} + c_3\,e^{(.65271\,x)}$$

Alternatively, we can first define the differential equation in (1) by entering the command

```
diffEq := diff(y(x),x$3)-3*diff(y(x),x$2)+y(x) = 0;
```

$$diffeq := \left(\frac{\partial^3}{\partial^3 x}y(x)\right) - 3\left(\frac{\partial^2}{\partial^2 x}y(x)\right) + y(x) = 0$$

and then ask for its exact symbolic solution:

```
soln := dsolve( diffEq, y(x) ):
```

This gives the complicated form with the complex exponents referred to in Eq. (3), but the floating point evaluation

```
soln := evalf(soln);
```

gives the approximate solution obtained above (though with the imaginary round-off errors still visible).

Using *Mathematica*

The characteristic equation in (2) is defined by

```
charEq =   x^3 - 3x^2 + 1 == 0;
```

The polynomial solve command

```
Solve[ charEq, r]
```

yields the complex "Cardan formula expressions" for the three roots exhibited previously, but the numerical solve command

```
soln = NSolve[ charEq, r ]
```

$\{\{r\text{->}-0.532089\}, \{r\text{->}\ 0.652704\}, \{r\text{->}\ 2.87939\}\}$

```
roots = r /.{{r->0.532089},{r->0.652704},{r->2.87939}}
```

$\{-0.532089,\ 0.652704, 2.87939\}$

gives the approximate roots of the characteristic equation (2) mentioned earlier. We can now assemble an approximate solution of the differential equation in (1):

```
c = {c1,c2,c3};
y == Apply[Plus, c Exp[roots x]]
```

$$y == c_1 e^{-0.532089x} + c_2 e^{0.652704x} + c_3 e^{2.97939x}$$

Alternatively, we can first define the differential equation in (1) by entering the command

```
diffEq = y'''[x] - 3 y''[x] + y[x] == 0
```

and then ask for its solution by means of the DE solve command

```
DSolve[diffEq, y[x], x]
```

$$\left\{\left\{y(x) \to c_1 e^{x\,\text{Root}(\#1^3-3\#1^2+1\&,1)} + c_2 e^{x\,\text{Root}(\#1^3-3\#1^2+1\&,2)} + c_3 e^{x\,\text{Root}(\#1^3-3\#1^2+1\&,3)}\right\}\right\}$$

This expresses the general solution in terms of the three symbolic roots of the characteristic equation, but the numerical evaluation

```
y[x] /. N[soln] // First
```

$$2.71828^{-0.532089x}c_1 + 2.71828^{0.652704x}c_2 + 2.71828^{2.97939x}c_3$$

of the result gives the approximate numeric form (4) of the solution — provided that we recognize the three appearances of the exponential base $e \approx 2.71828$.

Using MATLAB

We can work either in a purely numeric mode or in a symbolic mode. For a numerical approach, the characteristic polynomial in (2) is defined by the vector `[1 -3 0 1]` listing its coefficients in order of descending powers. Then the command

```
roots([1 -3 0 1])
ans =

    2.8794
    0.6527
   -0.5321
```

yields the three approximate characteristic roots that appear in the approximate general solution (4). The symbolic command

```
soln = solve('x^3 - 3*x^2 + 1 = 0')
```

yields "Cardan formula expressions" for these three roots similar to symbolic expressions exhibited originally. But then the command

```
numeric(soln)
ans =
   2.8794 + 0.0000i
  -0.5321 + 0.0000i
   0.6527 - 0.0000i
```

reproduces the approximate roots obtained previously. With a bit of work we can assemble the corresponding approximate solution:

```
syms C1 C2 C3 x
y1 = ['C1*e^',num2str(real(numeric(soln(1)))),'x'];
y2 = ['C2*e^',num2str(real(numeric(soln(2)))),'x'];
y3 = ['C3*e^',num2str(real(numeric(soln(3)))),'x'];
soln = [y1,'+',y2,'+',y3]
```

```
soln =
C1*e^2.8794x+C2*e^-0.53209x+C3*e^0.6527x
```

Alternatively, we can first define the differential equation in (1) by entering the symbolic command

```
diffeq = 'D3y - 3*D2y + y = 0'
diffeq =
D3y - 3*D2y + y = 0
```

and then ask for its solution by means of the command

```
y = dsolve(diffeq)
```

We can extract three independent particular solutions y_1, y_2, and y_3 from the resulting symbolic general solution by isolating the coefficients of the arbitrary constants C_1, C_2, and C_3:

```
y1 = subs(y,{C1,C2,C3},{1,0,0})
y1 =
exp((-299539413723655/562949953421312+
                    1/9007199254740992*i)*x)

y2 = subs(y,{C1,C2,C3},{0,1,0})
y2 =
exp((6483799150499469/2251799813685248-
                    1/9007199254740992*i)*x)

y3 = subs(y,{C1,C2,C3},{0,0,1})
y3 =
exp((1469757945450895/2251799813685248-
                    1/9007199254740992*i)*x)
```

Finally, we cut/paste/evaluate the rational fractions that appear as real parts in the exponents here, so as to verify that they correspond to the three approximate characteristic roots found previously.

```
-299539413723655/562949953421312
ans =
   -0.5321

6483799150499469/2251799813685248
ans =
    2.8794

1469757945450895/2251799813685248
ans =
    0.6527
```

Application 3.5
Automated Variation of Parameters

The method of variation of parameters — as described in the final part of Section 3.5 in the text — is readily implemented using a computer algebra system. In the paragraphs below we illustrate the use of *Maple*, *Mathematica*, and MATLAB in finding a particular solution of the differential equation

$$y'' + y = \tan x \tag{1}$$

of Example 11 in Section 3.5, to which the method of undetermined coefficients does not apply. In each case the initial commands serve to enter the two independent homogeneous solutions $y_1 = \cos x$ and $y_2 = \sin x$ and the nonhomogeneous term $f(x) = \tan x$ in Equation (1). The final commands implement the variation of parameters formula

$$y_p(x) = -y_1(x)\int \frac{y_2(x) f(x)}{W(x)}\,dx + y_2(x)\int \frac{y_1(x) f(x)}{W(x)}\,dx \tag{2}$$

To solve similarly another second-order linear differential equation

$$y'' + P(x)\,y' + Q(x)\,y = f(x) \tag{3}$$

whose complementary function $y_c(x) = c_1 y_1(x) + c_2 y_2(x)$ is known, we need only insert the corresponding definitions of $y_1(x)$, $y_2(x)$, and $f(x)$ in the initial commands. Find in this way the indicated particular solution $y_p(x)$ of each of the nonhomogeneous equations in Problems 1–6 below.

1. $y'' + y = 2\sin x$ $\quad y_p(x) = -x\cos x$

2. $y'' + y = 4x\sin x$ $\quad y_p(x) = x\sin x - x^2\cos x$

3. $y'' + y = 12x^2\sin x$ $\quad y_p(x) = 3x^2\sin x + \left(3x - 2x^3\right)\cos x$

4. $y'' - 2y' + 2y = 2e^x\sin x$ $\quad y_p(x) = -xe^x\cos x$

5. $y'' - 2y' + 2y = 4xe^x\sin x$ $\quad y_p(x) = e^x\left(x\sin x - x^2\cos x\right)$

6. $y'' - 2y' + 2y = 12x^2 e^x \sin x \qquad y_p(x) = e^x\left[3x^2 \sin x + \left(3x - 2x^3\right)\cos x\right]$

Using *Maple*

First we enter the independent complementary solutions

```
y1 := cos(x):
y2 := sin(x):
```

and the nonhomogeneous term

```
f := tan(x):
```

in Eq. (1). Then we calculate and simplify the Wronskian

```
W := y1*diff(y2,x) - y2*diff(y1,x):
W := simplify(W);
```

$$W := 1$$

of y_1 and y_2. It remains only to calculate the desired particular solution

```
yp := -y1*int(y2*f/W,x)+y2*int(y1*f/W,x):
yp := simplify(yp);
```

$$yp := -\cos(x)\ln\left(\frac{1+\sin(x)}{\cos(x)}\right)$$

using formula (2) above. Do you see that this result is equivalent to the particular solution

$$y_p = -(\cos x)\ln(\sec x + \tan x)$$

found in the text?

Using *Mathematica*

First we enter the independent complementary solutions

```
y1 = Cos[x];
y2 = Sin[x];
```

and the nonhomogeneous term

```
f = Tan[x];
```

in Eq. (1). Then we calculate and simplify the Wronskian

```
W  = y1*D[y2,x] - y2*D[y1,x];
W = Simplify[W]
1
```

of y_1 and y_2. It remains only to calculate the desired particular solution

```
yp = -y1*Integrate[y2*f/W,x] + y2*Integrate[y1*f/W,x];
yp = Simplify[yp]
```

$$-\cos(x)\left(\log\left(\cos\left(\frac{x}{2}\right)-\sin\left(\frac{x}{2}\right)\right)-\log\left(\cos\left(\frac{x}{2}\right)+\sin\left(\frac{x}{2}\right)\right)\right)$$

using formula (2) above. If you write this result in the form

$$y_p = -(\cos x)\ln\frac{\cos\frac{x}{2}-\sin\frac{x}{2}}{\cos\frac{x}{2}+\sin\frac{x}{2}}$$

and begin by multiplying numerator and denominator of the fraction by $\cos(x/2)-\sin(x/2)$, you should be able to use familiar trigonometric identities to show that the result is equivalent to the particular solution

$$y_p = -(\cos x)\ \ln(\sec x + \tan x)$$

found in the text.

Using MATLAB

First we enter the independent complementary solutions

```
syms x
y1 = cos(x);
y2 = sin(x);
```

and the nonhomogeneous term

```
f = tan(x)
```

in Eq. (1). Then we calculate and simplify the Wronskian

```
W = y1*diff(y2,x) - y2*diff(y1,x);
```

```
W = simplify(W)

W =
    1
```

of y_1 and y_2. It remains only to calculate the desired particular solution

```
yp = -y1*int(y2*f/W,x)+y2*int(y1*f/W,x);
yp = simplify(yp)
yp =
-cos(x)*log((1+sin(x))/cos(x))
```

using formula (2) above. Thus we find the particular solution

$$y_p(x) = -\cos(x)\ln\left(\frac{1+\sin(x)}{\cos(x)}\right)$$

Do you see that this result is equivalent to the particular solution

$$y_p = -(\cos x)\ \ln(\sec x + \tan x)$$

found in the text?

Application 3.6
Forced Vibrations and Resonance

Here we investigate forced vibrations of the mass-spring-dashpot system with equation

$$m x'' + c x' + k x \;=\; F(t) \qquad (1)$$

To simplify the notation, let's take $m = p^2$, $c = 2p$, and $k = p^2 q^2 + 1$ where $p, q > 0$. Then the complementary function of Eq. (1) is

$$x_c(t) \;=\; e^{-t/p}\left(c_1 \cos qt + c_2 \sin qt\right). \qquad (2)$$

We will take $p = 5$, $q = 3$ and thus investigate the transient and steady periodic solutions corresponding to

$$25x'' + 10x' + 226x \;=\; F(t), \quad x(0) = 0, \quad x'(0) = 0 \qquad (3)$$

with several illustrative possibilities for the external force $F(t)$. For your personal investigations to carry out similarly, you might select integers p and q with $6 \le p \le 9$ and $2 \le q \le 5$.

Investigation 1
With periodic external force $F(t) \;=\; 901\cos 3t$ the graph of the solution $x(t)$ of the resulting initial value problem in (3) is shown in Fig. 3.6.13 in the text. There we see the (transient plus steady periodic) solution

$$x(t) \;=\; \cos 3t + 30\sin 3t + e^{-t/5}\left[-\cos 3t - (451/15)\sin 3t\right]$$

rapidly "building up" to the steady periodic oscillation $x_{sp}(t) \;=\; \cos 3t + 30\sin 3t$.

Investigation 2
With damped oscillatory external force $F(t) \;=\; 900e^{-t/5}\cos 3t$ we have duplication with the complementary function in (2). The graph of the solution $x(t)$ of the resulting initial value problem in (3) is shown in Fig. 3.6.14 in the text. There we see the solution

$$x(t) \;=\; 6te^{-t/5}\sin 3t$$

oscillating up-and-down between the envelope curves $x = \pm 6t\,e^{-t/5}$. (Note the t-factor that betokens a resonance situation.)

Investigation 3

With damped oscillatory external force $F(t) = 2700t\,e^{-t/5}\cos 3t$ we have a still more complicated resonance situation. The graph of the solution $x(t)$ of the resulting initial value problem in (3) is shown in Fig. 3.6.15 in the text. There we see the solution

$$x(t) = e^{-t/5}\left[3t\cos 3t + \left(9t^2 - 1\right)\sin 3t\right]$$

oscillating up-and-down between the envelope curves $x = \pm e^{-t/5}\sqrt{(3t)^2 + (9t^2-1)^2}$.

Using *Maple*

With the damped oscillatory external force

```
F := 900*exp(-t/5)*cos(3*t):
```

of Investigation 2 we have the differential equation

```
de :=
25*diff(x(t),t,t)+10*diff(x(t),t)+226*x(t) = F;
```

$$de := 25\left(\frac{\partial^2}{\partial t^2}x(t)\right) + 10\left(\frac{\partial}{\partial t}x(t)\right) + 226\,x(t) = 900\,e^{\left(-\frac{1}{5}t\right)}\cos(3t)$$

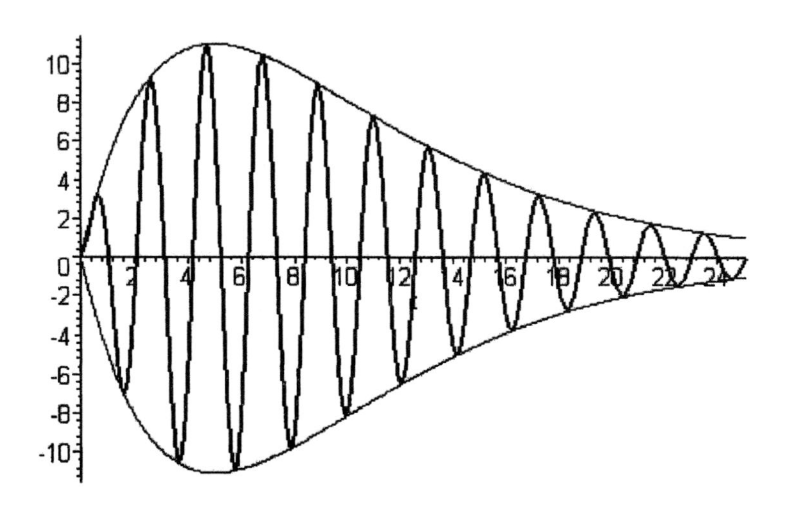

The solution of the initial value problem in (3) is then given by

```
dsolve({de,x(0)=0,D(x)(0)=0}, x(t)):
x := simplify(combine(rhs(%),trig));
```

$$x := 6e^{\left(-\frac{1}{5}t\right)}\sin(3t)\,t$$

Thus we have a damped oscillation with amplitude function

```
C :=  6*t*exp(-t/5):
```

The command

```
plot([x,C,-C],t=0..8*Pi);
```

then produces the figure shown on the preceding page.

Using *Mathematica*

With the damped oscillatory external force

```
F = 2700t Exp[-(t/5)] Cos[3t];
```

of Investigation 3 we have the differential equation

```
de =   25 x''[t] + 10 x'[t] + 226 x[t] == F
```

$$226\,x(t) + 10\,x'(t) + 25\,x''(t) == 2700\,e^{-t/5}t\cos(3t)$$

The solution of the initial value problem in (3) is then given by

```
soln =
DSolve[{de, x[0]==0, x'[0]==0}, x[t], t];

x = First[x[t] /. soln] // Simplify
```

$$e^{-t/5}\left(3t\cos(3t) + (9t^2-1)\sin(3t)\right)$$

Thus we have a damped oscillation with amplitude function

```
amp = Exp[-(t/5)] Sqrt[(3t)^2 + (9t^2-1)^2]
```

$$e^{-t/5}\sqrt{9t^2 + (9t^2-1)^2}$$

The command

```
Plot[{x, amp,-amp}, {t, 0,10 Pi}];
```

then produces the figure shown at the top of the next page.

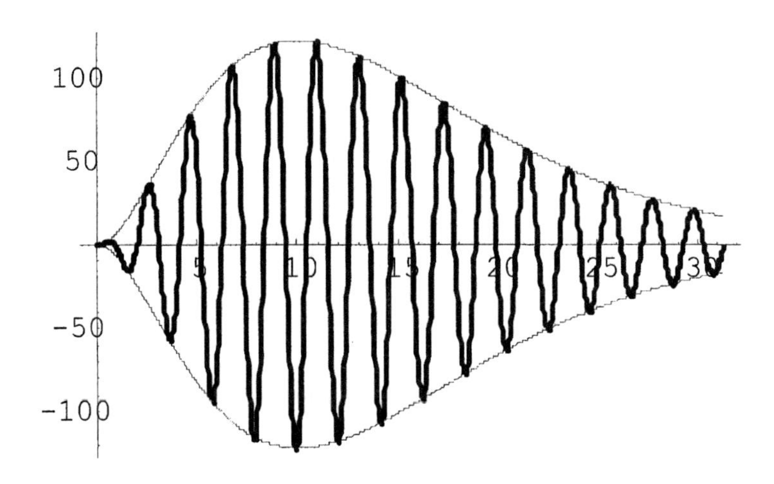

Using MATLAB

With the periodic external force $F(t) = 901\cos 3t$ of Investigation 1 our initial value problem is

$$25x'' + 10x' + 226x = 901\cos 3t, \quad x(0) = 0, \quad x'(0) = 0.$$

We proceed to solve this problem using MATLAB's symbolic **dsolve** function.

```
dsolve('25*D2x+10*Dx+226*x=901*cos(3*t)',
           'x(0)=0, Dx(0)=0');
x = simple(ans)
x =
cos(3*t)+30*sin(3*t)-exp(-1/5*t)*cos(3*t)-
     451/15*exp(-1/5*t)*sin(3*t)
```

We see that this particular solution is the sum of the steady periodic solution

```
t = 0 : pi/100 : 6*pi;
xsp = cos(3*t)+30*sin(3*t);
```

and the transient solution

```
xtr = -exp(-t/5).*(cos(3*t)+(451/15)*sin(3*t));
```

The plot commands

```
plot(t, xsp),
axis([0 6*pi -40 40])
hold on
plot(t, xsp + xtr)
```

finally produce the plot shown below. We see the (transient plus steady periodic) solution

$$x(t) = \cos 3t + 30\sin 3t + e^{-t/5}\left[-\cos 3t - (451/15)\sin 3t\right]$$

rapidly "building up" to the steady periodic oscillation $x_{sp}(t) = \cos 3t + 30\sin 3t$.

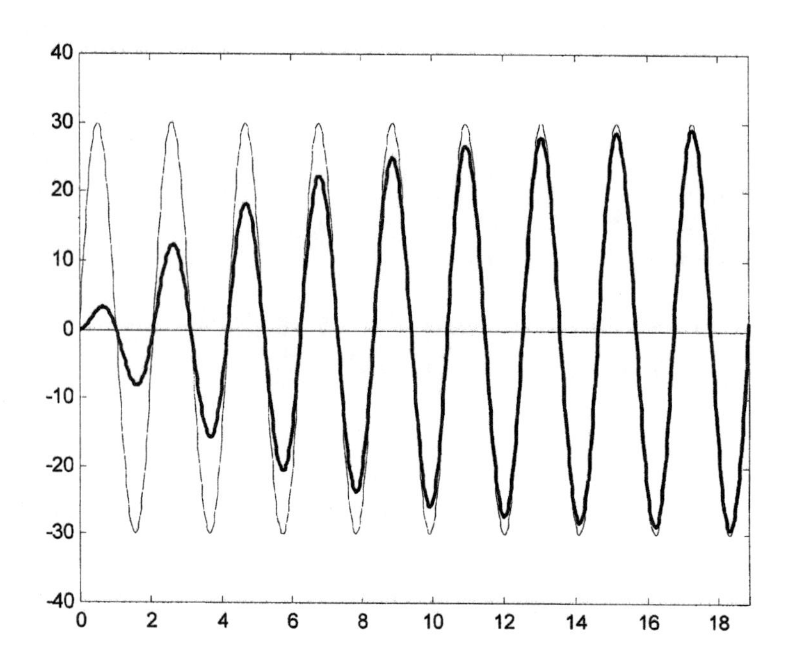

Chapter 4

Introduction to Systems of Differential Equations

Application 4.1
Kepler's Laws and Planetary Orbits

The Section 4.1 project in the text starts with Newton's inverse-square law of gravitation and outlines a derivation of the polar-coordinate formula

$$r(\theta) = \frac{L}{1+\varepsilon\cos(\theta-\alpha)} \tag{1}$$

describing an elliptical planetary orbit with eccentricity ε and semi-latus rectum L. The angle α is the planet's polar coordinate angle at perihelion — when it is closest to the sun. If the numerical values of L, ε, and α are known, then the ellipse can be plotted in rectangular coordinates by writing

$$x(t) = r(t)\cos t, \quad y(t) = r(t)\sin t, \quad 0 \le t \le 2\pi \tag{2}$$

(with parameter t in place of θ).

Here we want to use formulas (1) and (2) to plot some typical planetary orbits, starting with data found in a common source like a world almanac — where a planet's maximum and minimum distances r_M and r_m (respectively) from the sun typically are listed, but its semi-latus rectum is unlikely to be mentioned. But if we take $\alpha = 0$ in (1) then it should be clear that

$$r_M = \frac{L}{1-\varepsilon} \quad \text{and} \quad r_m = \frac{L}{1+\varepsilon}. \tag{3}$$

Upon equating values of L, these two equations are easily solved first for

$$\varepsilon = \frac{r_M - r_m}{r_M + r_m} \tag{4}$$

and then for

$$L = r_M(1-\varepsilon) = r_m(1+\varepsilon). \qquad (5)$$

The initial columns in the table below list the maximum and minimum distances from the sun (in astronomical units, where 1 AU = 93 million miles is the mean distance of the Earth from the sun) of the nine planets and Halley's comet. The last two columns list values of ε and L calculated using Eqs. (4) and (5).

Planet	r_M	r_m	ε	L
Mercury	0.467	0.308	0.2056	0.371
Venus	0.728	0.718	0.0067	0.723
Earth	1.017	0.983	0.0172	1.000
Mars	1.667	1.382	0.0935	1.511
Jupiter	5.452	4.953	0.0480	5.190
Saturn	10.081	9.015	0.0558	9.518
Uranus	19.997	17.949	0.0540	18.918
Neptune	30.341	29.682	0.0110	30.008
Pluto	48.940	29.639	0.2456	36.919
Halley	35.304	0.587	0.9673	1.155

In the paragraphs below we illustrate the use of *Maple*, *Mathematica*, and MATLAB to plot typical planetary orbits. You can try these and others.

Using *Maple*

To plot the orbit of the planet Mercury, we first enter its maximum and minimum distances from the sun.

```
r1 := 0.467:
r2 := 0.308:
```

Then we calculate its eccentricity and semi-latus rectum using Eqs (4) and (5).

```
e := (r1 - r2)/(r1 + r2):
L := r1*(1-e):
```

Now we can calculate its polar and rectangular coordinate functions using (1) and (2).

```
r := L/(1+e*cos(t)):
x := r*cos(t):
y := r*sin(t):
```

and finally plot the orbit (with $\alpha = 0$).

```
plot([x,y,t=0..2*Pi],
     view=[-0.5..0.5, -0.5..0.5],
     thickness = 2, color = red,
     scaling = constrained,
     title = "Mercury Orbit");
```

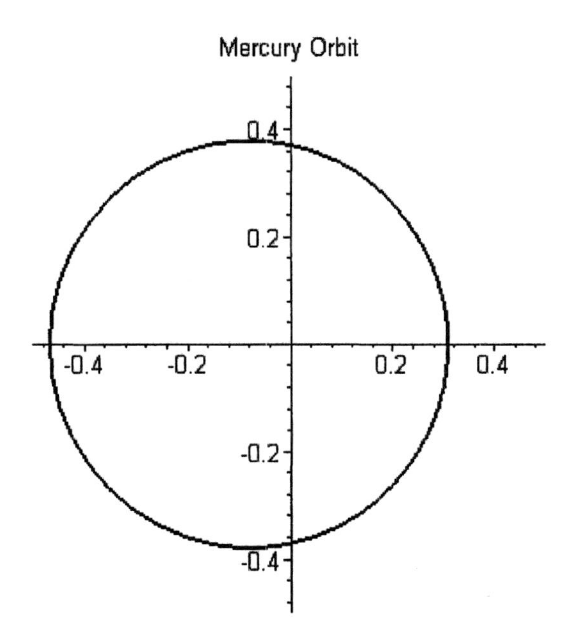

The option **scaling=constrained** insures equal scales on the *x*- and *y*-axes. We see that the elliptical orbit of Mercury actually looks quite circular. The non-uniformity of the motion of Mercury consists in the facts that this "circle" is off-center from the sun at the origin, and that its speed varies with its position on the orbit, with the planet moving fastest at perihelion and slowest at aphelion.

Using *Mathematica*

To plot the orbit of the planet Mercury, we first enter its maximum and minimum distances from the sun.

```
r1 := 0.467:
r2 := 0.308:
```

Then we calculate its eccentricity and semi-latus rectum using Eq. (4) and (5).

```
e := (r1 - r2)/(r1 + r2):
L := r1*(1-e):
```

Now we can calculate its polar and rectangular coordinate functions using (1) and (2).

```
r := L/(1+e*cos(t)):
x := r*cos(t):
y := r*sin(t):
```

and finally plot the orbit (with $\alpha = 0$).

```
ParametricPlot[{x, y}, {t, 0, 2*Pi},
   PlotRange -> {{-45, 5}, {-5, 5}},
   AspectRatio -> 0.2, AxesLabel -> {"x", "y"},
   PlotStyle -> {Thickness[0.0065],RGBColor[1,0, 0]},
        PlotLabel -> "Orbit of Halley's Comet"];
```

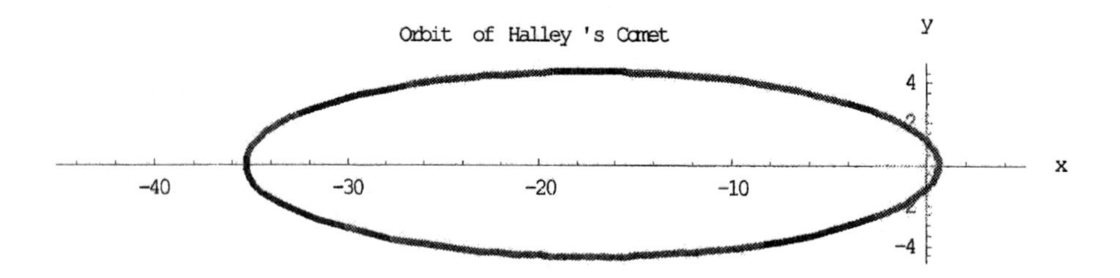

The option **AspectRatio->1** serves to ensure equal scales on the x- and y-axes. The plotted orbit certainly looks rather eccentric, though perhaps not so much as the actual eccentricity of $\varepsilon \approx 0.97$ might lead on to expect.

Using MATLAB

To plot the Earth's orbit, we first enter its maximum and minimum distances from the sun.

```
r1 = 1.017;
r2 = 0.983;
```

Then we calculate its eccentricity and semi-latus rectum using Eqs. (4) and (5).

```
e = (r1 - r2)/(r1 + r2);
L = r1*(1-e);
```

Now we can calculate its polar and rectangular coordinate functions using (1) and (2).

```
t = 0 : pi/100 : 2*pi;
r = L./(1+e*cos(t));
x = r.*cos(t);
y = r.*sin(t);
```

and finally plot the orbit (with $\alpha = 0$).

```
h = plot(x,y,'b');
set(h,'linewidth',2);
w = 0.025;  % to set viewing window
axis(w*[-r1 r1 -r1 r1]), axis square
hold on
plot(w*[-r1 r1],[0 0],'k')
plot([0 0],w*[-r1 r1;],'k')
title('Earth Orbit')
```

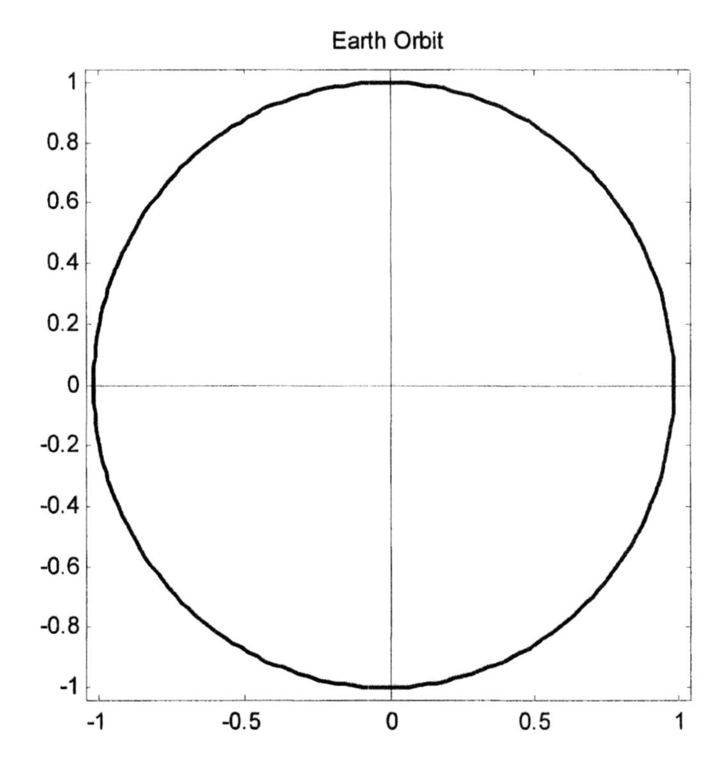

The fact that the Earth's orbit is elliptical is so engrained in modern minds that the seemingly circular appearance of the orbit (on any reasonable scale) may come as a surprise. We have plotted the orbit in the square viewing window $-1.025 \le x, y \le 1.025$ so that a careful examination of the figure will reveal that this "circular" orbit is visibly off-center from the sun at the origin. This fact, together with the non-uniformity of the Earth's speed in its orbit — it moves fastest at perihelion and slowest at aphelion — constitutes the actual ellipticity of the motion.

Application 4.2
Computer Algebra Solution of Systems

Computer algebra systems can be used to solve systems as well as single differential equations. The sections below illustrate the use of *Maple*, *Mathematica*, and MATLAB to solve symbolically the first-order system

$$x' = 4x - 3y, \quad y' = 6x - 7y \tag{1}$$

of Example 1 and the second-order mass-spring system

$$x'' + 3x - y = 0, \qquad y'' - 2x + 2y = 0 \tag{2}$$

of Example 3 in Section 4.2 of the text.

Use these examples as models for the computer algebra solution of several of the systems in Problems 1–20 and 39–46 in Section 4.2 (supplying initial conditions for the latter ones if you like). In each case, you should verify that the general solution obtained by computer is equivalent to the one obtained manually. Frequently the two presumably equivalent solutions will look different at first glance, and you will then need to explore the relation between the arbitrary constants in the computer algebra solution and the arbitrary constants in the manual solution.

Using *Maple*

To solve the system in (1) we need only define the differential equations

```
deq1 :=  diff(x(t),t) = 4*x(t) - 3*y(t);
deq2 :=  diff(y(t),t) = 6*x(t) - 7*y(t);
```

$$deq1 := \frac{\partial}{\partial x}x(t) = 4x(t) - 3y(t)$$

$$deq2 := \frac{\partial}{\partial x}y(t) = 6x(t) - 7y(t)$$

Then the command

```
dsolve( {deq1,deq2}, {x(t),y(t)} );
```

$$\{y(t) = \frac{9}{7}\_C1\,e^{(-5t)} - \frac{2}{7}\_C1\,e^{(2t)} + \frac{6}{7}\_C2\,e^{(2t)} - \frac{6}{7}\_C2\,e^{(-5t)},$$

$$x(t) = -\frac{3}{7}\_C1\,e^{(2t)} + \frac{3}{7}\_C1\,e^{(-5t)} - \frac{2}{7}\_C2\,e^{(-5t)} + \frac{9}{7}\_C2\,e^{(2t)}\}$$

yields (after a bit of simplification) the general solution

$$\begin{aligned} x(t) &= \tfrac{1}{7}(3c_1 - c_2)e^{-5t} + \tfrac{1}{7}(-3c_1 + 9c_2)e^{2t} \\ y(t) &= \tfrac{1}{7}(9c_1 - 6c_2)e^{-5t} + \tfrac{1}{7}(-2c_1 + 6c_2)e^{2t}. \end{aligned} \tag{3}$$

Is it clear that this result agrees with the general solution

$$x(t) = \tfrac{3}{2}b_1e^{2t} + \tfrac{1}{3}b_2e^{-5t}, \quad y(t) = b_1e^{2t} + b_2e^{-5t} \tag{4}$$

found in the text? What is the relation between the constants c_1, c_2 in (3) and the constants b_1, b_2 in (4)?

To find a particular solution we need only include the desired initial conditions in the **dsolve** command. Thus

```
dsolve({deq1,deq2, x(0)=2,y(0)=-1}, {x(t),y(t)});
```

$$\{y(t) = -3\,e^{(-5t)} + 2\,e^{(2t)}, x(t) = 3\,e^{(2t)} - e^{(-5t)}\}$$

solves the initial value problem of Example 1 in the text.

For the second-order mass-spring system in (2) we define the differential equations

```
deq3 := diff(x(t),t,t) + 3*x(t) -   y(t) = 0:
deq4 := diff(y(t),t,t) - 2*x(t) + 2*y(t) = 0:
```

and proceed to solve for the general solution.

```
dsolve( {deq3,deq4}, {x(t),y(t)} );
```

$$\{y(t) = \frac{2}{3}\_C1\cos(t)+\frac{1}{3}\_C1\cos(2t)+\frac{1}{6}\_C2\sin(2t)+\frac{2}{3}\_C2\sin(t)$$
$$-\frac{2}{3}\_C3\cos(2t)+\frac{2}{3}\_C3\cos(t)-\frac{1}{3}\_C4\sin(2t)+\frac{2}{3}\_C4\sin(t),$$
$$x(t) = -\frac{1}{3}\_C1\cos(2t)+\frac{1}{3}\_C1\cos(t)-\frac{1}{6}\_C2\sin(2t)+\frac{1}{3}\_C2\sin(t)$$
$$+\frac{1}{3}\_C3\cos(t)+\frac{2}{3}\_C3\cos(2t)+\frac{1}{3}\_C4\sin(2t)+\frac{1}{3}\_C4\sin(t)\}$$

Upon comparing the coefficients of cos(x), sin(x), cos($2x$), and sin($2x$) in these expressions for x and y, we see that this general solution is a linear combination of

- an oscillation with frequency 1 in which the two masses move synchronously with the amplitude of the second mass motion being twice the amplitude of the first, and

- an oscillation of frequency 2 in which the two masses move in opposite directions with the same amplitude of motion.

Hence it is equivalent to the general solution found in Example 3 of the text.

The two oscillations described above are more readily visible in the solution of the initial value problem

```
dsolve( {deq3,deq4, x(0)=0, y(0)=0,
        D(x)(0)=6, D(y)(0)=6}, {x(t),y(t)} );
```

$$\{x(t) = 4\sin t + \sin 2t, \; y(t) = 8\sin t - \sin 2t\}$$

Using *Mathematica*

To solve the system in (1) we need only define the differential equations

```
deq1 =    x'[t] == 4 x[t] - 3 y[t];
deq2 =    y'[t] == 6 x[t] - 7 y[t];
```

Then the command

```
DSolve[ {deq1,deq2}, {x[t],y[t]}, t ]
```

$$\{\{x(t) \to e^{-5t} c_1 + 3e^{2t} c_2, \; y(t) \to 3e^{-5t} c_1 + 2e^{2t} c_2\}\}$$

yields the general solution

$$x(t) = 3c_2e^{2t} + c_1e^{-5t}, \quad y(t) = 2c_2e^{2t} + 3c_1e^{-5t} \tag{5}$$

Compare this result with the general solution

$$x(t) = \tfrac{3}{2}b_1e^{2t} + \tfrac{1}{3}b_2e^{-5t}, \quad y(t) = b_1e^{2t} + b_2e^{-5t} \tag{6}$$

found in the text? What is the relation between the constants c_1, c_2 in (5) and the constants b_1, b_2 in (6)?

To find a particular solution we need only include the desired initial conditions in the **DSolve** command. Thus

```
DSolve[{deq1,deq2,x[0]==2,y[0]==-1}, {x[t],y[t]}, t];
```

$$\{\{x(t) \to e^{-5t}(-1+3e^{7t}),\ y(t) \to e^{-5t}(-3+2e^{7t})\}\}$$

solves the initial value problem of Example 1 in the text.

For the second-order mass-spring system in (2) we define the differential equations

```
deq3 =  x''[t] + 3*x[t] -    y[t] == 0;
deq4 =  y''[t] - 2*x[t] + 2*y[t] == 0;
```

The command

```
genSoln = DSolve[ {deq3,deq4}, {x[t],y[t]}, t ];
```

then yields $x(t)$ and $y(t)$ as rather miserable-looking linear combinations of the complex exponentials e^{it}, e^{-it}, e^{2it}, and e^{-2it} (try it for yourself and see). But we can convert these complex exponentials to trigonometric expressions by means of the commands

```
xg = First[x[t] /. genSoln] // ComplexExpand;
Collect[xg, {Cos[t],Sin[t],Cos[2t],Sin[2t]} ]
```

$$(i c_1 - i c_2)\cos(t) + (i c_4 - i c_3)\cos(2t) + (c_1 + c_2)\sin(t) + (-c_3 - c_4)\sin(2t)$$

```
yg = First[y[t] /. genSoln] // ComplexExpand;
Collect[yg, {Cos[t],Sin[t],Cos[2t],Sin[2t]} ]
```

$$(2i c_1 - 2i c_2)\cos(t) + (i c_3 - i c_4)\cos(2t) + (2c_1 + 2c_2)\sin(t) + (c_3 + c_4)\sin(2t)$$

Upon comparing the coefficients of $\cos(x)$, $\sin(x)$, $\cos(2x)$, and $\sin(2x)$ in these expressions for $x(t)$ and $y(t)$, we see that this general solution is a linear combination of

- an oscillation with frequency 1 in which the two masses move synchronously with the amplitude of the second mass motion being twice the amplitude of the first, and
- an oscillation of frequency 2 in which the two masses move in opposite directions with the same amplitude of motion.

Hence it is equivalent to the general solution found in Example 3 of the text.

The two oscillations described above are more readily visible in the solution of the initial value problem

```
partSoln =
```

```
DSolve[{deq3, deq4, x[0]==0,y[0]==0,
        x'[0]==6,y'[0]==6}, {x[t], y[t]}, t];

xp = First[x[t] /. partSoln] //
          ComplexExpand // Simplify
```

$4\sin(t)+\sin(2t)$

```
yp = First[y[t] /. partSoln] //
          ComplexExpand // Simplify
```

$8\sin(t)-\sin(2t)$

Using MATLAB

To solve the system in (1) we need only define the differential equations

```
deq1 =  'Dx = 4*x - 3*y';
deq2 =  'Dy = 6*x - 7*y';
```

Then the command

```
[x,y] = dsolve(deq1,deq2)
x =
  -2/7*C1*exp(-5*t)+9/7*C1*exp(2*t)-
   3/7*C2*exp(2*t)+3/7*C2*exp(-5*t)
y =
   6/7*C1*exp(2*t)-6/7*C1*exp(-5*t)+
   9/7*C2*exp(-5*t)-2/7*C2*exp(2*t)
```

yields (after a bit of simplification) the general solution

$$\begin{aligned} x(t) &= \tfrac{1}{7}(3c_1 - 2c_2)e^{-5t} + \tfrac{1}{7}(-3c_1 + 9c_2)e^{2t} \\ y(t) &= \tfrac{1}{7}(9c_1 - 6c_2)e^{-5t} + \tfrac{1}{7}(-2c_1 + 6c_2)e^{2t}. \end{aligned} \tag{7}$$

Is it clear that this result agrees with the general solution

$$x(t) = \tfrac{3}{2}b_1e^{2t} + \tfrac{1}{3}b_2e^{-5t}, \quad y(t) = b_1e^{2t} + b_2e^{-5t} \tag{8}$$

found in the text? What is the relation between the constants c_1, c_2 in (7) and the constants b_1, b_2 in (8)?

To find a particular solution we need only include the desired initial conditions in the **dsolve** command. Thus

```
[x,y] = dsolve(deq1,deq2,'x(0)=2','y(0)=-1')

x =
  -exp(-5*t)+3*exp(2*t)
y =
   2*exp(2*t)-3*exp(-5*t)
```

solves the initial value problem of Example 1 in the text.

For the second-order mass-spring system in (2) we define the differential equations

```
deq3 = 'D2x + 3*x - y = 0';
deq4 = 'D2y - 2*x + 2*y = 0';
```

and proceed to solve for the general solution.

```
[x,y] = dsolve(deq3,deq4)
x =
   1/3*C1*cos(t)+2/3*C1*cos(2*t)+1/3*C2*sin(2*t)
  +1/3*C2*sin(t)-1/3*C3*cos(2*t)+1/3*C3*cos(t)
  -1/6*C4*sin(2*t)+1/3*C4*sin(t)
y =
  -2/3*C1*cos(2*t)+2/3*C1*cos(t)-1/3*C2*sin(2*t)
  +2/3*C2*sin(t)+2/3*C3*cos(t)+1/3*C3*cos(2*t)
  +1/6*C4*sin(2*t)+2/3*C4*sin(t)
```

Upon comparing the coefficients of $\cos(x)$, $\sin(x)$, $\cos(2x)$, and $\sin(2x)$ in these expressions for x and y, we see that this general solution is a linear combination of

- an oscillation with frequency 1 in which the two masses move synchronously with the amplitude of the second mass motion being twice the amplitude of the first, and
- an oscillation of frequency 2 in which the two masses move in opposite directions with the same amplitude of motion.

Hence it is equivalent to the general solution found in Example 3 of the text.

The two oscillations described above are more readily visible in the solution of the initial value problem

```
[x,y] = dsolve(deq3,deq4, 'x(0)=0, y(0)=0,
                     Dx(0)=6, Dy(0)=6')

x =
   sin(2*t)+4*sin(t)
y =
  -sin(2*t)+8*sin(t)
```

Further Investigation

Now that we have found the general solution of the system in (1), we can use *Mathematica* to display the solution graphically. (*Maple* and MATLAB users can substitute equivalent commands.) After the commands

```
deq1 = x'[t] -> 4x[t] - 3y[t];
deq2 = y'[t] -> 6x[t] - 7y[t];
```

shown above, the following commands will draw in the *xy*-plane the particular solutions of the system that pass through the points (i, j), where i and j range independently from -3 to 3 (this is called a *phase portrait* for the system):

```
genSoln = DSolve[{deq1, deq2}, {x[t], y[t]}, t]
partSolns = Table[genSoln/.{C[1] -> i, C[2] -> j},
                  {i, -3, 3}, {j, -3, 3}]
partSolns = Flatten[partSolns, 2]
curves = ParametricPlot[Evaluate[{x[t],
              y[t]}/.partSolns], {t, -2, 2},
              PlotPoints -> 100, AspectRatio -> 1,
              PlotRange -> {{-3, 3}, {-3, 3}} ];
```

Can you see from the phase portrait some features of the solutions of the system that you would not have guessed from the algebraic solution in (5)?

We can take things further by adding a *direction field* to the solution curves; this will tell us the direction of increasing *t* along each solution curve. The idea is that at each point of the *xy*-plane, the vector $\langle dx/dt, dy/dt \rangle$ is tangent to the solution curve passing through that point, and in fact points in the direction of increasing *t*. First we load in a *Mathematica* "package", then draw the direction field, and finally combine it with the phase portrait:

```
<<Graphics`PlotField`
field = PlotVectorField[{4x - 3y, 6x - 7y},
          {x, -3, 3}, {y, -3, 3},
          ScaleFunction -> (.5&)];
Show[curves, field];
```

Now that you have the "big picture" in front of you, answer these questions:

- What general shape do all of the solution curves seem to have? How would you describe them?
- How many solution curves aren't actually "curved"?

- Suppose you were to close your eyes, choose a point (x, y) at random, and follow the solution curve passing through that point. Where will you likely wind up?
- Suppose you wanted to start at a point (x, y) other than the origin and wind up (as $t \to \infty$) at the origin—from what points (x, y) could you do this?
- Finally: Would any of the above questions been easy to answer without the phase portrait and direction field we have drawn here?

Feel free to peek ahead at Section 6.1 (as well as the rest of Chapter 6) to see some of the many other things one can learn from phase portraits and direction fields.

Application 4.3A

The Runge-Kutta Method for 2-Dimensional Systems

Figure 4.3.11 in the text lists TI-85 and BASIC versions of the program **RK2DIM** that implements the Runge-Kutta iteration

$$
\begin{aligned}
k_1 &= f(t_n, x_n, y_n) \\
l_1 &= g(t_n, x_n, y_n) \\
k_2 &= f(t_n + \tfrac{1}{2}h, x_n + \tfrac{1}{2}h\, k_1, y_n + \tfrac{1}{2}h\, l_1) \\
l_2 &= g(t_n + \tfrac{1}{2}h, x_n + \tfrac{1}{2}h\, k_1, y_n + \tfrac{1}{2}h\, l_1) \\
k_3 &= f(t_n + \tfrac{1}{2}h, x_n + \tfrac{1}{2}h\, k_2, y_n + \tfrac{1}{2}h\, l_2) \\
l_3 &= g(t_n + \tfrac{1}{2}h, x_n + \tfrac{1}{2}h\, k_2, y_n + \tfrac{1}{2}h\, l_2) \\
k_4 &= f(t_n + h, x_n + h\, k_3, y_n + h\, l_3) \\
l_4 &= f(t_n + h, x_n + h\, k_3, y_n + h\, l_3) \\
k &= \tfrac{1}{6}(k_1 + 2k_2 + 2k_3 + k_4), \\
l &= \tfrac{1}{6}(l_1 + 2l_2 + 2l_3 + l_4) \\
x_{n+1} &= x_n + h\, k \\
y_{n+1} &= y_n + h\, l \\
t_{n+1} &= t_n + h
\end{aligned}
\tag{1}
$$

for the two dimensional system

$$
\begin{aligned}
x' &= f(t, x, y), \\
y' &= g(t, x, y)
\end{aligned}
\tag{2}
$$

You should note that **RK2DIM** closely parallels the one-dimensional Runge-Kutta program listed in Fig. 2.6.11, with a single line to calculate one value or slope there replaced (where appropriate) with two lines here to calculate a *pair* of x- and y-values or slopes. Note also that the notation used is essentially that of Equations. (13) and (14) in Section 4.3.

The sections below illustrate the use of *Maple*, *Mathematica*, and MATLAB to implement the Runge-Kutta iteration in (1), and apply it to the initial value problem

$$
\begin{aligned}
x' &= -\pi\, y, \qquad & x(0) &= 1 \\
y' &= \pi\, x, \qquad & y(0) &= 0
\end{aligned}
\tag{3}
$$

whose exact solution is given by

$$x(t) = \cos(\pi t), \qquad y(t) = \sin(\pi t).$$

This exercise should prepare you for the following investigations.

Investigation A

Let a be the largest and b the smallest nonzero digit of your student ID number. Then use the Runge-Kutta method to approximate the solution of the initial value problem defined by

$$x' = -a\,y, \qquad y' = b\,x$$

with $x(0)$ being the next smallest digit and $y(0)$ the next largest digit of your ID. Also use your system's "dsolve" function to find the exact symbolic solution. Finally, compare your approximate solution points with the corresponding exact solution points.

Investigation B

Suppose that *you* jump from an airplane at an initial altitude of 10,000 ft, and that your *downward* position $x(t)$ and velocity $y(t)$ (in ft/sec after t seconds) satisfy the initial value problem

$$\begin{aligned} x' &= y, & x(0) &= 0 \\ (W/g)y' &= W - (0.0015)(100y + y^2), & y(0) &= 0 \end{aligned}$$

where W denotes *your* weight in pounds and $g = 32$ ft/sec^2. Use the Runge-Kutta method to solve numerically for $x(t)$ and $y(t)$ during your first 15 to 20 seconds of free fall, with successive step sizes h and $h/2$ small enough to get results consistent to 2 decimal places. How far have you free-fallen after 10 seconds? After 20 seconds? Use your results (for the first 20 seconds) to determine the total time required for your descent to the ground.

Investigation C

Your spacecraft is traveling at constant velocity V, approaching a distant earth-like planet with mass M and radius R. When activated, your deceleration system provides a constant thrust T until impact with the planet's surface. During the deceleration period, your distance $x(t)$ from the planet's center satisfies the differential equation

$$\frac{d^2x}{dt^2} = T - \frac{GM}{x^2}$$

where $G \approx 6.6726 \times 10^{-11}\ \text{N} \cdot (\text{m/kg})^2$ is the usual gravitational constant. Your question is — At what altitude above the planet's surface should your deceleration system be activated in order to achieve a soft touchdown? For a reasonable problem, you can take

$$
\begin{aligned}
M &= 5.97 \times 10^{24} \text{ kg} \\
R &= 6.38 \times 10^{6} \text{ m} \\
V &= p \times 10^{4} \text{ km/hr} \\
T &= g + q \text{ m/s}^2,
\end{aligned}
$$

where $g = GM/R^2$ is the planet's surface gravitational acceleration, while p is the smallest nonzero digit and q the next-smallest digit in your student ID number. Find the "ignition altitude" accurate to the nearest meter, and the resulting "descent time" accurate to the nearest tenth of a second.

Using *Maple*

To apply the Runge-Kutta method to the initial value problem in (3), we let

```
Digits := 6:
pi := evalf(Pi)   # numerical value of Pi
```

and define the right-hand side functions in our two differential equations:

```
f := (t,x,y) -> -pi*y:
g := (t,x,y) ->  pi*x:
```

Suppose that we want to approximate the solution functions $x = \cos(\pi t)$ and $y = \sin(\pi t)$ on the interval $[0, 1/2]$ corresponding to angular arguments between 0 and $\pi/2$. To approximate the solution with initial values $x(t_0) = x_0$, $y(t_0) = y_0$ on the interval $[t_0, t_f]$, we enter first the initial values (and final t-value)

```
t0 := 0:
x0 := 1:        y0 := 0:
tf := 0.5:
```

and then the desired number n of steps, the interval at which we want to print results, and the resulting step size h.

```
n := 12:                  # number of subintervals
m := 2:                   # to print every mth step
h := evalf((tf - t0)/n):  # step size
```

After we initialize the values of t, x, and y,

```
t := t0:      x := x0:      y := y0:
```

the Runge-Kutta method is implemented by the **for** loop below, which carries out the iteration in (1) n times in successtion to take n steps across the interval from $t = t_0$ to $t = t_f$. It simplies the printing to define in advance the "formatting string"

```
fmt := `%10.0f %10.4f %10.4f \n`:
```

where the notation **%w.df** specifies printing the corresponding value in a "field" w spaces wide and with d decimal places. Then Runge-Kutta loop is then

```
for i from 1 to n do
   k1 := f(t,x,y):
   l1 := g(t,x,y):                         # left-hand slopes
   k2 := f(t+h/2,x+h*k1/2,y+h*l1/2):
   l2 := g(t+h/2,x+h*k1/2,y+h*l1/2):       # 1st midpt slopes
   k3 := f(t+h/2,x+h*k2/2,y+h*l2/2):
   l3 := g(t+h/2,x+h*k2/2,y+h*l2/2):       # 2nd midpt slopes
   k4 := f(t+h,x+h*k3,y+h*l3):
   l4 := g(t+h,x+h*k3,y+h*l3):             # right-hand slopes
   k := (k1+2*k2+2*k3+k4)/6:               # average x-slope
   l := (l1+2*l2+2*l3+l4)/6:               # average y-slope
   x := x + h*k:                           # update x
   y := y + h*l:                           # update y
   t := t + h:                             # update t
   if trunc(i/m) =i/m then                 # display each
        printf(fmt,180*t,x,y) fi;          #  mth value
   od:

        15          .9659       .2588
        30          .8660       .5000
        45          .7071       .7071
        60          .5000       .8660
        75          .2588       .9659
        90          .0000      1.0000
```

Note that we have arranged to print angles in *degrees* in the first column of output. The second and third columns give the corresponding approximate values of the cosine and sine functions. Noting the familiar values $\cos 60° = \sin 30° = \frac{1}{2}$, $\cos 30° = \sin 60° = \frac{1}{2}\sqrt{3} \approx 0.8660$, and $\cos 45° = \sin 45° = \frac{1}{2}\sqrt{2} \approx 0.7071$, we see that the Runge-Kutta method with only $n = 12$ subintervals has provided 4 decimal places of accuracy on the whole range from 0° to 90°.

If only the final endpoint result is wanted explicitly, then the print command can be removed from the loop and executed immediately following it (just as we did with the Euler loop in Project 2.4). For a different initial value problem, we need only enter the appropriate functions $f(x, y)$ and $g(x, y)$ for our new differential equations and the desired initial and final values in the initial commands above, then re-execute the subsequent ones.

Using *Mathematica*

To apply the Runge-Kutta method to the initial value problem in (3), we let

```
pi = N[Pi];            (* numerical value of Pi *)
```

and define the right-hand side functions in our two differential equations:

```
f[t_,x_,y_] := -pi*y
g[t_,x_,y_] :=  pi*x
```

Suppose that we want to approximate the solution functions $x = \cos(\pi t)$ and $y = \sin(\pi t)$ on the interval $[0, 1/2]$ corresponding to angular arguments between 0 and $\pi/2$. To approximate the solution with initial values $x(t_0) = x_0$, $y(t_0) = y_0$ on the interval $[t_0, t_f]$, we enter first the initial values (and final t-value)

```
t0 = 0;    x0 = 1;    y0 = 0;    tf = 0.5;
```

and then the desired number n of steps, the interval at which we want to print results, and the resulting step size h.

```
n = 12;              (* number of subintervals   *)
m = 2;               (* to print every mth step  *)
h = (tf - t0)/n;     (* step size                *)
```

After we initialize the values of t, x, and y,

```
t = t0;     x = x0;     y = y0;
```

the Runge-Kutta method itself is implemented by the **Do** loop below, which carries out the iteration in (1) n times in successtion to take n steps across the interval from $t = t_0$ to $t = t_f$.

```
Do[
   k1 = f[t,x,y];
   l1 = g[t,x,y];                      (* left-hand slopes *)
   k2 = f[t+h/2,x+h*k1/2,y+h*l1/2];
   l2 = g[t+h/2,x+h*k1/2,y+h*l1/2]; (* 1st midpt slopes *)
   k3 = f[t+h/2,x+h*k2/2,y+h*l2/2];
   l3 = g[t+h/2,x+h*k2/2,y+h*l2/2]; (* 2nd midpt slopes *)
   k4 = f[t+h,x+h*k3,y+h*l3];
   l4 = g[t+h,x+h*k3,y+h*l3];       (* right-hand slopes *)
```

```
k   = (k1+2*k2+2*k3+k4)/6;        (* average x-slope   *)
l   = (l1+2*l2+2*l3+l4)/6;        (* average y-slope   *)
x   = x + h*k;                    (* update x   *)
y   = y + h*l;                    (* update y   *)
t   = t + h;                      (* update t   *)
If[ Floor[i/m] == i/m,
    Print[180*t,PaddedForm[x,{10,4}],
    PaddedForm[y,{10,4}]]],
                         (* display each mth value *)
 {i,1,n} ]
```

```
15.       0.9659      0.2588
30.       0.8660      0.5000
45.       0.7071      0.7071
60.       0.5000      0.8660
75.       0.2588      0.9659
90.       0.0000      1.0000
```

Note that we have arranged to print angles in *degrees* in the first column of output. The second and third columns give the corresponding approximate values of the cosine and sine functions. The **PaddedForm[x,{10,4}]** specification formats a result by printing it in a "field" 10 spaces wide and with 4 decimal places. Noting the familiar values $\cos 60° = \sin 30° = \frac{1}{2}$, $\cos 30° = \sin 60° = \frac{1}{2}\sqrt{3} \approx 0.8660$, and $\cos 45° = \sin 45° = \frac{1}{2}\sqrt{2} \approx 0.7071$, we see that the Runge-Kutta method with only $n = 12$ subintervals has provided 4 decimal places of accuracy on the whole range from 0° to 90°.

If only the final endpoint result is wanted explicitly, then the print command can be removed from the loop and executed immediately following it (just as we did with the Euler loop in Project 2.4). For a different initial value problem, we need only enter the appropriate functions $f(x,y)$ and $g(x,y)$ for our new differential equations and the desired initial and final values in the initial commands above, then re-execute the subsequent ones.

Using MATLAB

To apply the Runge-Kutta method to the initial value problem in (3), we begin by defining the right-hand side functions $f(t,x,y)$ and $g(t,x,y)$ in our two differential equations:

```
f = inline('-pi*y','t','x','y');
g = inline('pi*x','t','x','y');
```

Suppose that we want to approximate the solution functions $x = \cos(\pi t)$ and $y = \sin(\pi t)$ on the interval [0, 1/2] corresponding to angular arguments between 0 and $\pi/2$. To approximate the solution with initial values $x(t_0) = x_0$, $y(t_0) = y_0$ on the interval $[t_0, t_f]$, we enter first the initial values (and final t-value)

```
t0 = 0;
x0 = 1;
y0 = 0;
tf = 0.5;
```

and then the desired number n of steps, the interval at which we want to print results, and the resulting step size h.

```
n = 12;                % number of subintervals
m =  2;                % to print every mth step
h = (tf - t0)/n;       % step size
```

After we initialize the values of t, x, and y,

```
t = t0;
x = x0;
y = y0;
```

and the first line of our desired table of values,

```
result = [t0,x0,y0];
result =
     0     1     0
```

the Runge-Kutta method itself is implemented by the **for** loop below, which carries out the iteration in (1) n times in successtion to take n steps across the interval from $t = t_0$ to $t = t_f$.

```
for i = 1:n,
    k1 = f(t,x,y);
    l1 = g(t,x,y);                       % left-hand slopes
    k2 = f(t+h/2,x+h*k1/2,y+h*l1/2);
    l2 = g(t+h/2,x+h*k1/2,y+h*l1/2); % 1st midpt slopes
    k3 = f(t+h/2,x+h*k2/2,y+h*l2/2);
    l3 = g(t+h/2,x+h*k2/2,y+h*l2/2); % 2nd midpt slopes
    k4 = f(t+h,x+h*k3,y+h*l3);
    l4 = g(t+h,x+h*k3,y+h*l3);          % right-hand slopes
    k = (k1+2*k2+2*k3+k4)/6;            % average x-slope
    l = (l1+2*l2+2*l3+l4)/6;            % average y-slope
    x = x + h*k;                     % Euler step to update x
    y = y + h*l;                     % Euler step to update y
```

```
   t = t + h;                    % update t
   if floor(i/m) == i/m,
      result = [result;[180*t,x,y]];
      end                        % adjoin new row of data
   end
```

Then the command

```
result
result =
         0    1.0000         0
   15.0000    0.9659    0.2588
   30.0000    0.8660    0.5000
   45.0000    0.7071    0.7071
   60.0000    0.5000    0.8660
   75.0000    0.2588    0.9659
   90.0000    0.0000    1.0000
```

displays our table of Runge-Kutta results.

Note that we have arranged to print angles in *degrees* in the first column of output. The second and third columns give the corresponding approximate values of the cosine and sine functions. Noting the familiar values $\cos 60° = \sin 30° = \frac{1}{2}$, $\cos 30° = \sin 60° = \frac{1}{2}\sqrt{3} \approx 0.8660$, and $\cos 45° = \sin 45° = \frac{1}{2}\sqrt{2} \approx 0.7071$, we see that the Runge-Kutta method with only $n = 12$ subintervals has provided 4 decimal places of accuracy on the whole range from 0° to 90°.

For a different initial value problem, we need only re-define the functions $f(x, y)$ and $g(x, y)$ corresponding to our new differential equations and the desired initial and final values in the initial commands above, then re-execute the subsequent ones.

Application 4.3B
Comets and Spacecraft

The investigations outlined here are intended as applications of the more sophisticated numerical DE solvers that are "built into" computing systems such as *Maple*, *Mathematica*, and MATLAB (as opposed to the *ad hoc* Runge-Kutta methods of the previous project.) We illustrate these high-precision variable step size solvers by applying them to analyze the elliptical orbit of a satellite — a comet, planet, or spacecraft — around a primary (planet or sun) of mass M. If the attracting primary is located at the origin in xyz-space, then the satellite's position functions $x(t)$, $y(t)$, and $z(t)$ satisfy Newton's inverse-square law differential equations

$$\frac{d^2x}{dt^2} = -\frac{\mu x}{r^3}, \qquad \frac{d^2y}{dt^2} = -\frac{\mu y}{r^3}, \qquad \frac{d^2z}{dt^2} = -\frac{\mu z}{r^3} \tag{1}$$

where $\mu = GM$ (G being the gravitational constant) and $r = \sqrt{x^2+y^2+z^2}$.

Investigation A

Consider a satellite in elliptical orbit around a planet, and suppose that physical units are so chosen that $\mu = 1$. If the orbit lies in the xy-plane so $z(t) \equiv 0$, the Eqs. (1) reduce to

$$\frac{d^2x}{dt^2} = -\frac{x}{r^3}, \qquad \frac{d^2y}{dt^2} = -\frac{y}{r^3}. \tag{2}$$

Let T denote the period of revolution of the satellite in its orbit. Kepler's third law says that the *square* of T is proportional to the *cube* of the major semiaxis a of its elliptical orbit. In particular, if $\mu = 1$, then

$$T^2 = 4\pi^2 a^3. \tag{3}$$

(See Section 12.6 of Edwards and Penney, *Calculus: Early Transcendentals*, 7th ed. (Prentice Hall, 2008).) If the satellite's x- and y-components of velocity, $x_2 = x' = x_1'$ and $y_2 = y' = y_1'$, are introduced, then the system in (2) translates into the system

$$\begin{aligned} x_1' &= x_2, & x_2' &= -\frac{\mu x_1}{r^3} \\ y_1' &= y_2, & y_2' &= -\frac{\mu y_1}{r^3} \end{aligned} \tag{4}$$

of four first-order equations with $r = \sqrt{x_1^2 + y_1^2}$.

(i) Solve Eqs. (2) or (4) numerically with the initial conditions

$$x(0) = 1, \qquad y(0) = 0, \qquad x'(0) = 0, \qquad y'(0) = 1$$

that correspond theoretically to a circular orbit of radius $a = 1$, in which case Eq. (3) gives $T = 2\pi$. Are your numerical results consistent with this fact?

(ii) Now solve the system numerically with the initial conditions

$$x(0) = 1, \qquad y(0) = 0, \qquad x'(0) = 0, \qquad y'(0) = \tfrac{1}{2}\sqrt{6}$$

that correspond theoretically to an elliptical orbit with major semiaxis $a = 2$, so Eq. (3) gives $T = 4\pi\sqrt{2}$. Do your numerical results agree with this?

(iii) Investigate what happens when both the x-component and the y-component of the initial velocity are nonzero.

Investigation B (Halley's Comet)

Halley's comet last reached perihelion (its point of closest approach to the sun at the origin) on February 9, 1986. Its position and velocity components at this time were

$$\mathbf{p}_0 = (0.325514, -0.459460, 0.166229) \quad \text{and}$$

$$\mathbf{v}_0 = (-9.096111, -6.916686, -1.305721)$$

(respectively) with position in AU (Astronomical Units, the unit of distance being equal to the major semiaxis of the earth's orbit about the sun) and time in years. In this unit system, its 3-dimensional equations of motion are as in (1) with $\mu = 4\pi^2$. Then solve Eqs. (1) numerically to verify the appearance of the yz-projection of the orbit of Halley's comet shown in Fig. 4.3.13 in the text. Plot the xy- and xz-projections also.

Figure 4.3.14 in the text shows the graph of the distance $r(t)$ of Halley's comet from the sun. Inspection of this graph indicates that Halley's comet reaches a maximum distance (at aphelion) of about 35 AU in a bit less than 40 years, and returns to perihelion after about three-quarters of a century. The closer look in Fig. 4.3.15 indicates that the period of revolution of Halley's comet is about 76 years. Use your numerical solution to refine these observations. What is your best estimate of the calendar date of the comet's next perihelion passage?

Investigation C (Your Own Comet)

Lucky you! The night before your birthday in 1997 you set up your telescope on nearby mountaintop. It was a clear night, and at 12:30 am you spotted a new comet. After repeating the observation on successive nights, you were able to calculate its solar system coordinates $\mathbf{p}_0 = (x_0, y_0, z_0)$ and its velocity vector $\mathbf{v}_0 = (vx_0, vy_0, vz_0)$ on that first night. Using this information, determine this comet's

- perihelion (point nearest sun) and aphelion (farthest from sun),
- its velocity at perihelion and at aphelion,
- its period of revolution about the sun, and
- its next two dates of perihelion passage.

Using length-time units of AU and earth years, the comet's equations of motion are given in (1) with $\mu = 4\pi^2$. For your personal comet, start with random initial position and velocity vectors with the same order of magnitude as those of Halley's comet. Repeat the random selection of initial position and velocity vectors, if necessary, until you get a nice-looking eccentric orbit that goes well outside the earth's orbit (like real comets do).

Using *Maple*

Let's consider a comet orbiting the sun with initial position and velocity vectors

```
p0 := {x(0)=0.2, y(0)=0.4, z(0)=0.2};
v0 := {D(x)(0)=5, D(y)(0)=-7, D(z)(0)=9};
```

at perihelion. For convenience, we combine these initial conditions in the single set

```
inits := p0 union v0;
```

of equations. The comet's equations of motion in (1) with $\mu = 4\pi^2$ are entered as

```
r :=   t->sqrt(x(t)^2 + y(t)^2 + z(t)^2);

de1 := diff(x(t),t$2)  =  -4*Pi^2*x(t)/r(t)^3:
de2 := diff(y(t),t$2)  =  -4*Pi^2*y(t)/r(t)^3:
de3 := diff(z(t),t$2)  =  -4*Pi^2*z(t)/r(t)^3:

deqs := {de1,de2,de3}:
```

The comet's *x*-, *y*-, and *z*-position functions then satisfy the combined set

```
eqs := deqs union inits:
```

of three second-order differential equations and six initial conditions, which we proceed to solve numerically.

```
soln := dsolve(eqs, {x(t),y(t),z(t)}, type=numeric);
```

soln := **proc**(*rkf45_x*) ... **end**

We use the resulting numerical procedure **soln** to plot the *yz*-projection of the comet's orbit for the first 20 years:

```
with(plots):
odeplot(soln, [y(t),z(t)], 0..20, numpoints=1000);
```

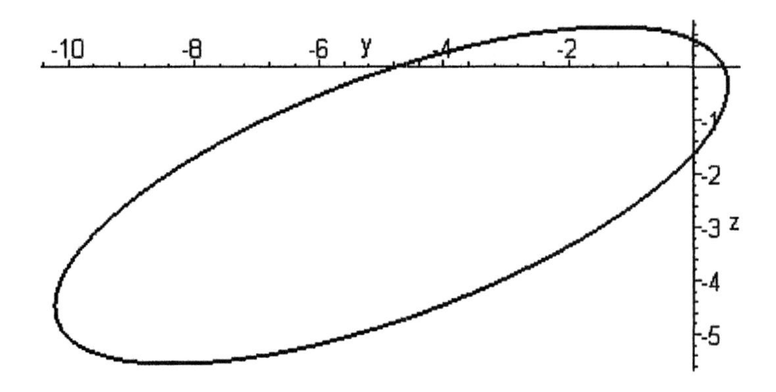

This orbit certainly looks like an ellipse. To investigate the comet's motion on its orbit, we plot its distance r from the sun as a function of t.

```
odeplot(soln, [t,r(t)], 0..20, numpoints=1000);
```

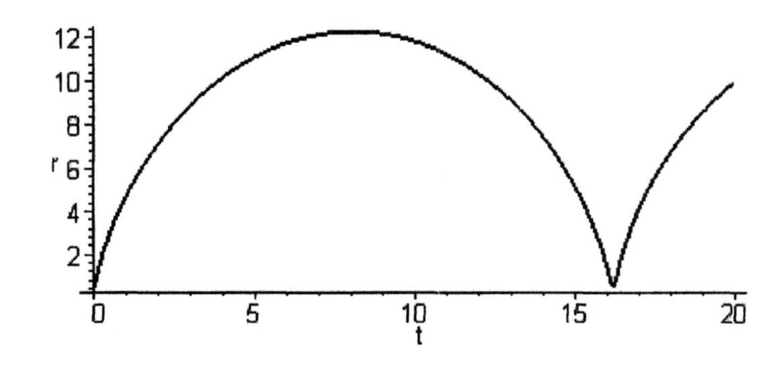

The comet appears to reach aphelion after about 8 years, and to return to perihelion after about 16 years. Zooming in on the aphelion,

```
odeplot(soln,[t,r(t)], 8.0..8.2);
```

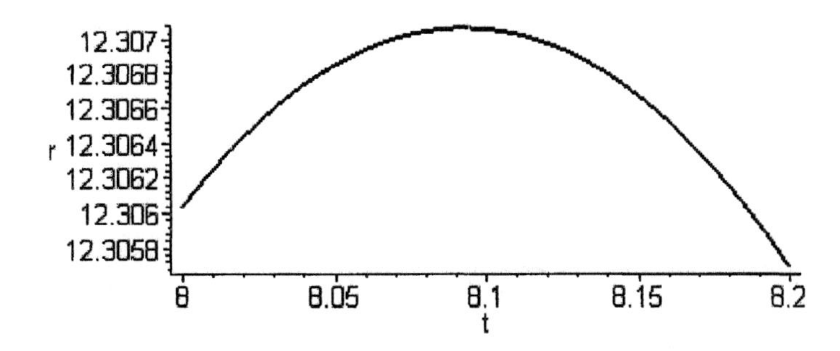

we see that the comet reaches a maximal distance from the sun of about 12.31 AU after about 8.09 years. Zooming in on the perihelion,

```
odeplot(soln, [t,r(t)], 16.1..16.3);
```

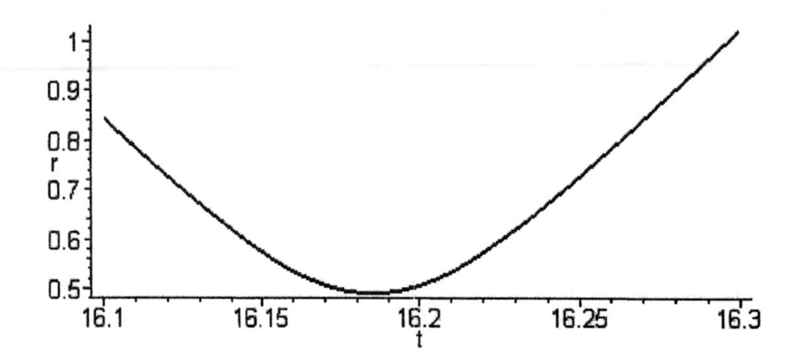

we see that the comet appears to return to a minimal distance of about 0.49 AU from the sun after about 16.18 years.

Using *Mathematica*

Let's consider a comet orbiting the sun with initial position and velocity vectors

```
p0 = {x[0]==0.2, y[0]==0.4, z[0]==0.2}
v0 = {x'[0]==5, y'[0]==-7, z'[0]==9}
```

at perihelion. For convenience, we combine these initial conditions in the single set

```
inits = Union[p0,v0]
```

of equations. The comet's equations of motion in (1) with $\mu = 4\pi^2$ are entered as

```
r[t_] = Sqrt[x[t]^2 + y[t]^2 + z[t]^2]

de1 =     x''[t] == -4 Pi^2 x[t]/r[t]^3;
de2 =     y''[t] == -4 Pi^2 y[t]/r[t]^3;
de3 =     z''[t] == -4 Pi^2 z[t]/r[t]^3;

deqs = {de1,de2,de3}
```

The comet's x-, y-, and z-position functions then satisfy the combined set

```
eqs = Union[deqs, inits]
```

of three second-order differential equations and six initial conditions, which we proceed to solve numerically.

```
soln = NDSolve[eqs, {x, y, z}, {t, 0, 20}]
```

```
{{x -> InterpolatingFunction[{{0.,20.}}, <>],
  y -> InterpolatingFunction[{{0.,20.}}, <>],
  z -> InterpolatingFunction[{{0.,20.}}, <>]}}
```

The result **soln** is a list of three numerical "interpolating functions"

```
x = First[x /. soln];
y = First[y /. soln];
z = First[z /. soln];
```

that we can use to plot the *yz*-projection of the comet's orbit for the first 20 years:

```
ParametricPlot[Evaluate[{y[t],z[t]}], {t,0,20}]
```

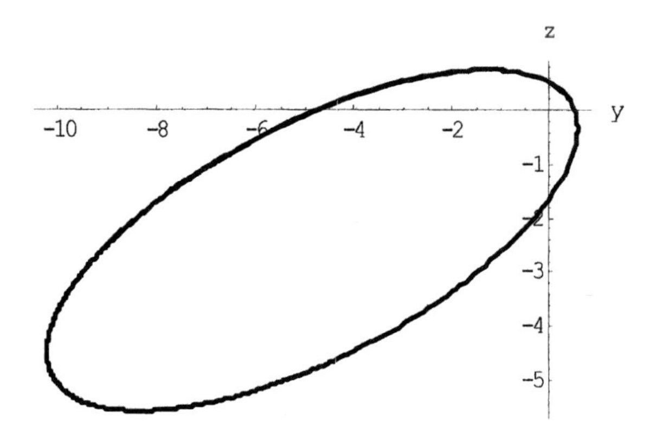

This certainly looks like an ellipse. To investigate the comet's motion on this orbit, we plot its distance r from the sun as a function of t.

```
Plot[Evaluate[r[t]], {t, 0, 20}]
```

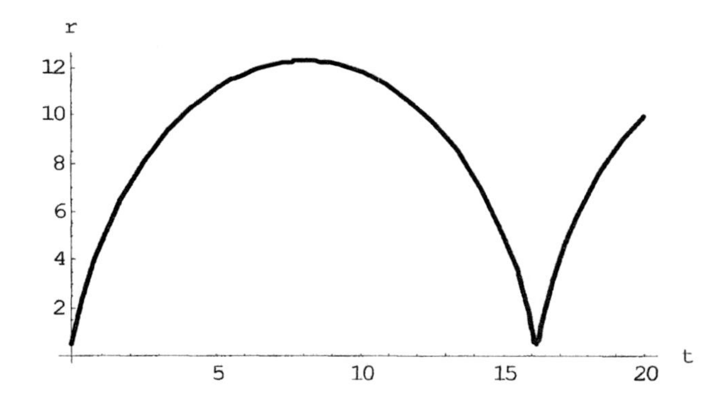

The comet appears to reach aphelion after about 8 years, and to return to perihelion after about 16 years. Zooming in on the aphelion,

```
Plot[Evaluate[r[t]], {t, 8.05, 8.15},
     PlotRange -> {12.3062, 12.3068}]
```

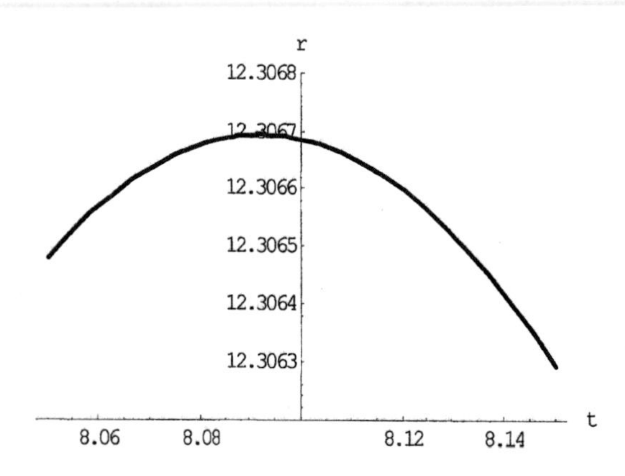

we see that the comet reaches a maximal distance from the sun of about 12.31 AU after about 8.09 years. Zooming in on the perihelion,

```
Plot[Evaluate[r[t]], {t, 16.15, 16.25},
     PlotRange -> {0.45, 0.55},
```

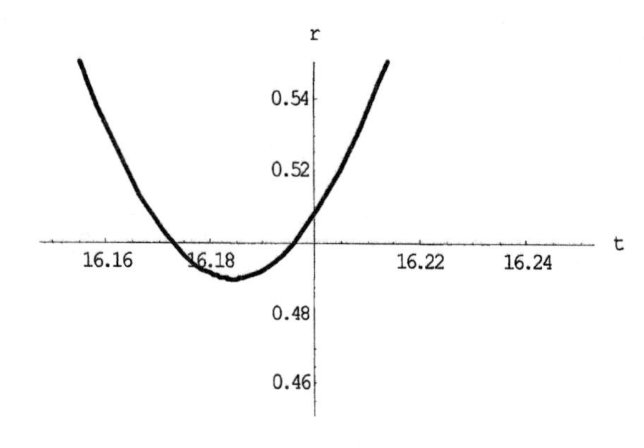

we see that the comet appears to return to a minimal distance of about 0.49 AU from the sun after about 16.18 years.

Using MATLAB

Let's consider a comet orbiting the sun with initial position and velocity column vectors

```
r0 = [0.2; 0.4; 0.2];
v0 = [5; -7; 9];
```

at perihelion. We combine these initial values into the single 6-component vector

```
inits = [p0; v0];
```

The following MATLAB function saved as **ypcomet.m** serves to define the comet's equations of motion in (1) with $\mu = 4\pi^2$.

```
function  yp = ypcomet(t,y)
yp = y;
vx = y(4);   vy = y(5);   vz = y(6);% velocity comps
x  = y(1);   z  = y(3);   y  = y(2);% coordinates
r  = sqrt( x*x + y*y + z*z );       % radius
r3  = r*r*r;                        % r-cubed
k = 4*pi^2;                         % for AU-yr units
yp(1)  =  vx;
yp(2)  =  vy;
yp(3)  =  vz;
yp(4)  =  -k*x/r3;
yp(5)  =  -k*y/r3;
yp(6)  =  -k*z/r3;
```

We proceed to solve these differential equations numerically with the given initial conditions.

```
options = odeset('reltol',1e-6);      % error tolerance
tspan = 0 : 0.01 : 20;  % from t=0 to t=20 with dt=0.01

[t,y] = ode45('ypcomet',0:0.01:20, inits, options);
```

Here **t** is the vector of times and **y** is a matrix whose first 3 column vectors give the corresponding position coordinates of the comet. We need only plot the second and third of these vectors against each other to see the *yz*-projection of the comet's orbit for the first 20 years.

```
plot(y(:,2),y(:,3)),
axis([-12 2 -12 2]), axis square
```

The resulting orbit (at the top of the next page) certainly looks like an ellipse.

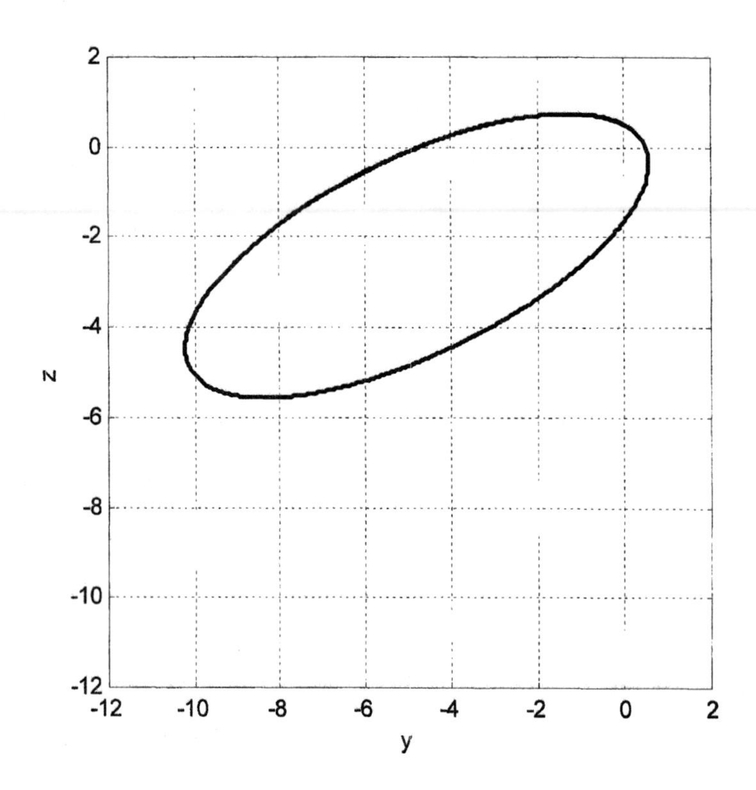

To investigate the comet's motion on this orbit, we plot its distance r from the sun as a function of t.

```
r = sqrt(y(:,1).^2 + y(:,2).^2 + y(:,3).^2);
plot(t, r)
```

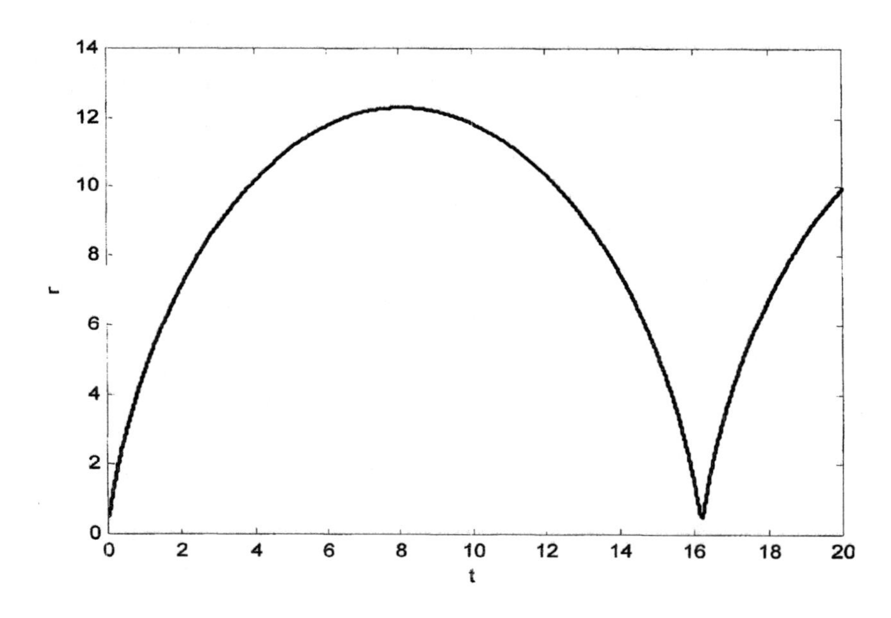

The comet appears to reach aphelion after about 8 years, and to return to perihelion after about 16 years. We can zoom in on the aphelion with the command

```
axis([8.05 8.15 12.3065 12.3075]), grid on
```

and see (in the figure below) that the comet reaches a maximal distance from the sun of about 12.31 AU after about 8.09 years.

We zoom in on the perihelion with the command

```
axis([16.1 16.3 0.4 0.8]), grid on
```

and see (in the figure at the top of the next page) that it appears to return to a minimal distance of about 0.49 AU from the sun after about 16.18 years.

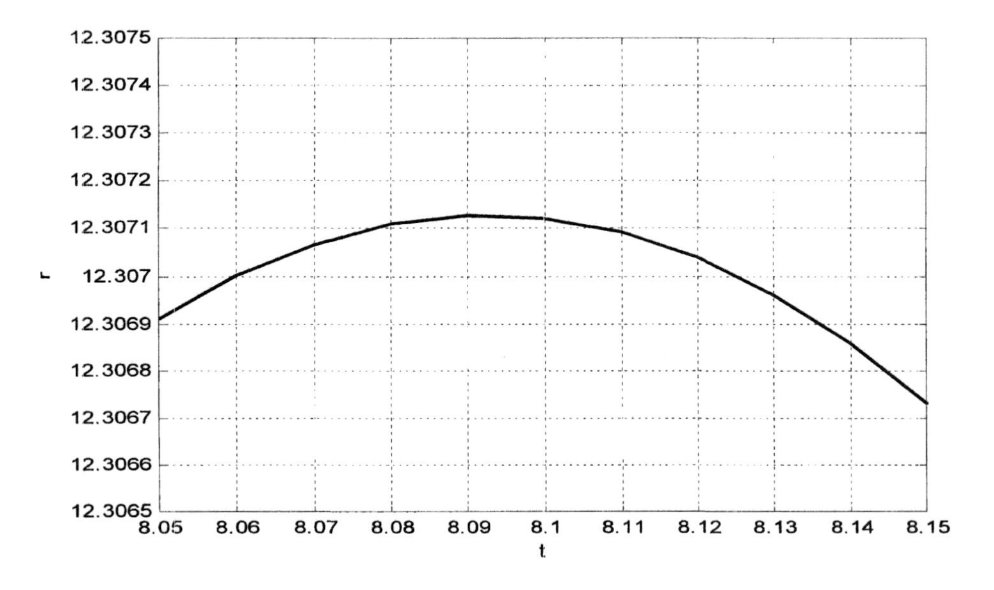

Zooming in on the comet's aphelion

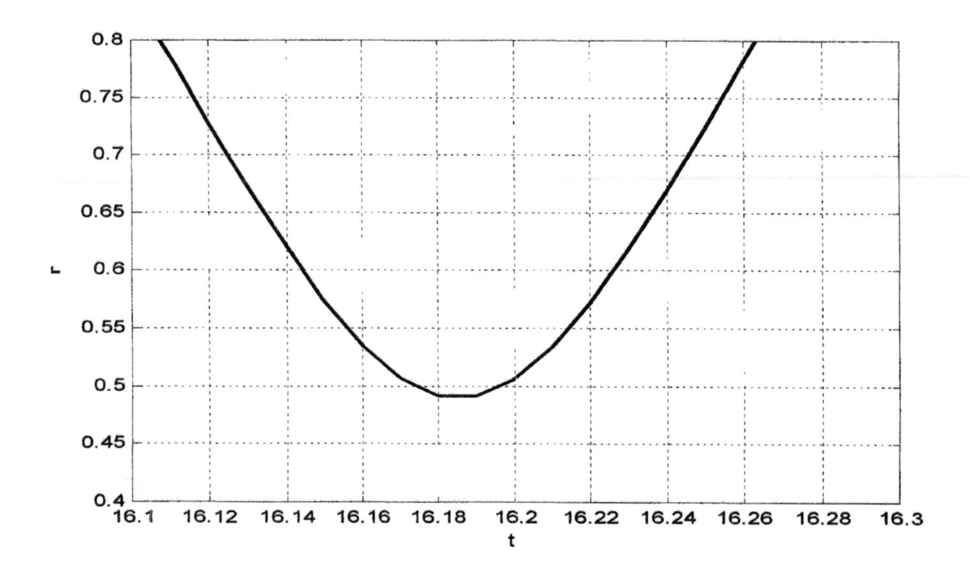

Zooming in on the comet's perihelion

Earth-Moon Satellite Orbits

We consider finally an Apollo satellite in orbit about the Earth E and the Moon M. Figure 4.3.8 in the text shows an x_1x_2-coordinate system whose origin lies at the center of mass of the Earth and the Moon, and which rotates at the rate of one revolution per "moon month" of approximately $\tau = 27.32$ days, so the Earth and Moon remain fixed in their positions on the x_1-axis. If we take as unit distance the distance between the Earth and Moon centers, then their coordinates are $E(-\mu, 0)$ and $M(1-\mu, 0)$, where $\mu = m_M/(m_E + m_M)$ in terms of the Earth mass m_E and the Moon mass m_M. If we take the total mass $m_E + m_M$ as the unit of mass and $\tau/2\pi \approx 4.348$ days as the unit of time, then the gravitational constant has value $G = 1$, and the equations of motion of the satellite position $S(x_1, x_2)$ are

$$x_1'' = x_1 + 2x_2' - \frac{(1-\mu)(x_1+\mu)}{r_E^3} - \frac{\mu(x_1-1+\mu)}{r_M^3} \tag{1a}$$

and

$$x_2'' = x_2 - 2x_1' - \frac{(1-\mu)x_2}{r_E^3} - \frac{\mu x_2}{r_M^3} \tag{1b}$$

where $r_E = \sqrt{(x_1+\mu)^2 + x_2^2}$ and $r_M = \sqrt{(x_1+\mu-1)^2 + x_2^2}$ denote the satellite's distance to the Earth and Moon, respectively. The initial two terms on the right-hand side of each equation in (1) result from the rotation of the coordinate system. In the system of units described here, the lunar mass is approximately $\mu = m_M = 0.012277471$. The second-order system in (1) can be converted to a first-order system by substituting

$$x_1' = x_3, \qquad x_2' = x_4$$

so

$$x_3' = x_1'', \qquad x_4' = x_2''. \tag{2}$$

This system is defined in the MATLAB function

```
function  yp = ypmoon(t,y)

m1 = 0.012277471;                 % mass of moon
m2 = 0.987722529;                 % mass of earth

r1 = norm([y(1)+m1, y(2)]);       % Distance to the earth
r2 = norm([y(1)-m2, y(2)]);       % Distance to the moon

yp = [ y(3); y(4); 0; 0 ];        % Column 4-vector

yp(3)  = y(1)+2*y(4)-m2*(y(1)+m1)/r1^3-m1*(y(1)-m2)/r2^3;

yp(4)  = y(2)-2*y(3)  - m2*y(2)/r1^3  - m1*y(2)/r2^3;
```

Suppose that the satellite initially is in a clockwise circular orbit of radius about 1500 miles about the Moon. At its farthest point from the Earth $(x_1 = 0.994)$ it is "launched" into Earth-Moon orbit with initial velocity v_0. We then want to solve the system in (2) — with the right-hand functions in (1) substituted for x_1'' and x_2'' — with the initial conditions

$$x_1(0) = 0.994, \qquad x_2(0) = 0, \qquad x_3(0) = 0, \qquad x_4(0) = -v_0. \tag{3}$$

In the system of units used here, the unit of velocity is approximately 2289 miles per hour. Some initial conditions and final times of particular interest are defined by the function

```
function  [tf,y0]  = mooninit(k)

% Initial conditions for k-looped Apollo orbit
```

```
if  k == 2,
   tf = 5.436795439260;
   y0 = [ 0.994 0 0 -2.113898796695 ]';

elseif  k == 3,
   tf = 11.124340337266;
   y0 = [ 0.994 0 0 -2.031732629557 ]';

elseif  k == 4,
   tf = 17.065216560158;
   y0 = [ 0.994 0 0 -2.001585106379 ]';
   end
```

The first two components of **y0** are the coordinates of the initial position, and the last two components are the components of the initial velocity; **tf** is then the time required to complete one orbit. The cases $k = 3$ and $k = 4$ yield Figures 4.3.9 and 4.3.10 (respectively) in the text. The following commands (with $k = 2$) yield the figure shown on the next page, and illustrate how such figures are plotted.

```
[tf,y0] = mooninit(2);
options = odeset('RelTol',1e-9,'AbsTol',1e-12);
[t,y] = ode45('ypmoon', [0,tf], y0, options);
plot(y(:,1), y(:,2));
axis([-1.5 1.3 -1.4 1.4]), axis square
```

The small relative and absolute error tolerances are needed to insure that the orbit closes smoothly when the satellite returns to its initial position.

You might like to try the values $k = 3$ and $k = 4$ to generate the analogous 3- and 4-looped orbits. A more substantial project would be to search empirically for initial velocities yielding periodic orbits with more than 4 loops.

Further Investigations

See the Application 4.3C page at the web site **www.prenhall.com/edwards** for additional investigations of comets, satellites, and trajectories of baseballs with air resistance (as in Example 4 of Section 4.3 in the text).

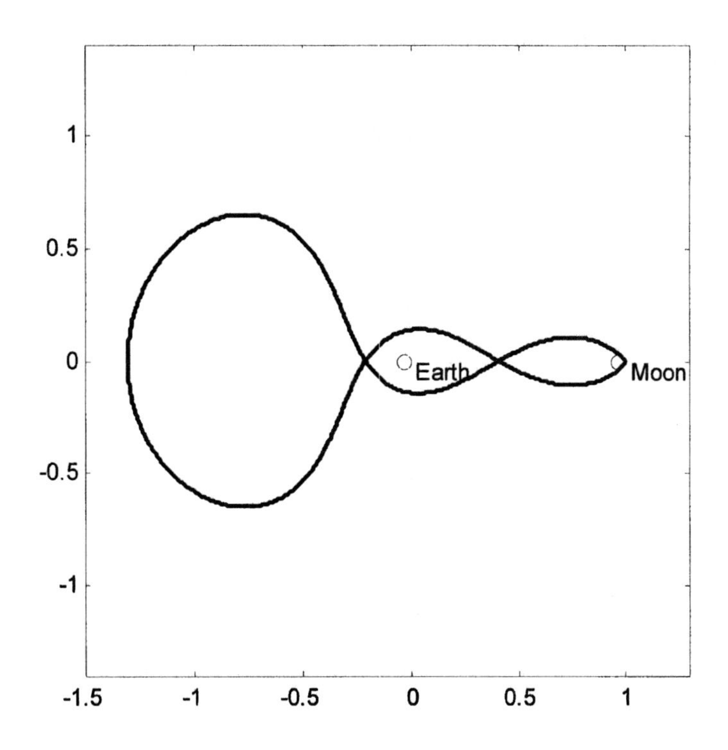

1
0.5
0
-0.5
-1
-1.5
-1
-0.5
0
0.5
1
Earth
Moon

Chapter 5

Linear Systems of Differential Equations

Application 5.1
Automatic Solution of Linear Systems

Calculations with numerical matrices of order greater than 3 are most frequently carried out with the aid of calculators or computers. For example, in Example 8 of Section 5.1 in the text we needed to solve the linear system

$$\begin{aligned} 2c_1 + 2c_2 + 2c_3 &= 0 \\ 2c_1 \qquad - 2c_3 &= 2 \\ c_1 - c_2 + c_3 &= 6. \end{aligned} \tag{1}$$

Writing this system in matrix notation $\mathbf{AC} = \mathbf{B}$ and working with a TI-86 calculator, for instance, we would store the 3×3 coefficient matrix $\mathbf{A}$ and the column vector $\mathbf{B}$ with the commands

```
[[2,2,2] [2,0,-2] [1,-1,1]] → A
[[0] [2] [6]] → B
```

in which each matrix and each of its successive rows in enclosed within square brackets. Then we calculate and store the solution $\mathbf{C} = \mathbf{A}^{-1}\mathbf{B}$ with the command

```
A⁻¹*B  →  C
```

which yields the result

```
[[ 2]
 [-3]
 [ 1]]
```

Thus $c_1 = 2$, $c_2 = -3$, and $c_3 = 1$, as we found using elementary row operations in Example 8 in the text.

Matrix notations and calculations in most computer systems are similar. In the sections below, we illustrate how to enter matrices and perform simple matrix calculations using *Maple*, *Mathematica*, and MATLAB. The "investigations" that follow are simple exercises to familiarize you with automatic matrix computation. Applications will appear in the remaining projects in this chapter.

Investigation A

To practice simple matrix algebra with whatever calculator or computer system is available, you might begin by defining a square matrix $\mathbf{A}$ of your own selection. Then calculate its inverse matrix $\mathbf{B}$ and check that the matrix product $\mathbf{AB}$ is the identity matrix. Do this with several square matrices of different dimensions.

Investigation B

Use matrix algebra as indicated above (for the computations of Example 8) to solve Problems 31 through 40 of Section 5.1 in the text.

Investigation C

An interesting $n \times n$ matrix is the **Hilbert matrix** $\mathbf{H}_n$ whose *ij*th element (in the *i*th row and *j*th column) is $1/(i+j-1)$. For instance, the 4×4 Hilbert matrix is

$$\mathbf{H}_4 = \begin{bmatrix} 1 & \frac{1}{2} & \frac{1}{3} & \frac{1}{4} \\ \frac{1}{2} & \frac{1}{3} & \frac{1}{4} & \frac{1}{5} \\ \frac{1}{3} & \frac{1}{4} & \frac{1}{5} & \frac{1}{6} \\ \frac{1}{4} & \frac{1}{5} & \frac{1}{6} & \frac{1}{7} \end{bmatrix}. \qquad (2)$$

Set up the Hilbert matrices of orders $n = 3, 4, 5, 6, \ldots,$ and calculate their inverse matrices. The results will look more interesting than you might anticipate.

(1) Show that the system $\mathbf{H}_3\mathbf{x} = \mathbf{b}$ has the exact solution $\mathbf{x} = [1 \;\; 1 \;\; 1]^T$ if

$$\mathbf{b} = [11/6 \;\; 13/12 \;\; 47/60]^T \approx [1.83333 \;\; 1.08333 \;\; 0.78333]^T,$$

whereas it has the exact solution $\mathbf{x} = [0.6 \;\; 2.4 \;\; 0]^T$ if $\mathbf{b} = [1.8 \;\; 1.1 \;\; 0.8]^T$. Thus linear systems with coefficient matrix $\mathbf{H}_3$ can be quite "unstable" with respect to roundoff errors in the constant vector $\mathbf{b}$. (This example is essentially the one given by Steven H. Weintraub on page 324 of the April 1986 issue of the *American Mathematical Monthly*.)

(2) Show that the system $\mathbf{H}_4\mathbf{x} = \mathbf{b}$ has the exact solution $\mathbf{x} = [1 \;\; 1 \;\; 1 \;\; 1]^T$ if

$$\mathbf{b} = \left[\tfrac{25}{12} \;\; \tfrac{77}{60} \;\; \tfrac{57}{60} \;\; \tfrac{319}{420}\right]^T \approx [2.08333 \;\; 1.28333 \;\; 0.95000 \;\; 0.75952]^T,$$

whereas it has the exact solution $\mathbf{x} = [1.28 \;\; -1.8 \;\; 7.2 \;\; -2.8]^T$ if

$$\mathbf{b} = [2.08 \ 1.28 \ 0.95 \ 0.76]^T.$$

Using *Maple*

In order to carry out matrix computations the *Maple* **linalg** package must first be loaded,

```
with(linalg):
```

A particular matrix can be entered either with a command of the form

```
A := matrix( 3,3, [2,2,2, 2,0,-2, 1,-1,1]);
```

$$A := \begin{bmatrix} 2 & 2 & 2 \\ 2 & 0 & -2 \\ 1 & -1 & 1 \end{bmatrix}$$

where the first two arguments prescribe the numbers of rows and columns and the third argument is a vector listing the elements of **A** row-by-row, or with one of the form

```
A := matrix( [[2,2,2], [2,0,-2], [1,-1,1]] );
```

in which the individual row vectors of **A** are prescribed. The inverse matrix $\mathbf{A}^{-1}$ is then calculated with the command

```
B := inverse(A);
```

$$B := \begin{bmatrix} \frac{1}{8} & \frac{1}{4} & \frac{1}{4} \\ \frac{1}{4} & 0 & \frac{-1}{2} \\ \frac{1}{8} & \frac{-1}{4} & \frac{1}{4} \end{bmatrix}$$

The function **evalm** (for **eval**uate **m**atrices) is used for matrix multiplication, with **&*** denoting the matrix multiplication operator itself. Thus we can verify that **B** actually is the inverse matrix of **A** with the calculation

```
evalm( B &* A );
```

$$\begin{bmatrix} 1 & 0 & 0 \\ 0 & 1 & 0 \\ 0 & 0 & 1 \end{bmatrix}$$

Having defined the right-hand side constant (column) vector

```
b := matrix( 3,1, [0,2,6]);
```

$$b := \begin{bmatrix} 0 \\ 2 \\ 6 \end{bmatrix}$$

in (1), we can then calculate the solution vector $\mathbf{c} = \mathbf{A}^{-1}\mathbf{b} = \mathbf{Bb}$ with the command

```
c := evalm( B &* b );
```

$$c := \begin{bmatrix} 2 \\ -3 \\ 1 \end{bmatrix}$$

For higher-dimensional linear systems the computation of the inverse matrix is not so efficient as immediate application of *Maple*'s **`linsolve`** function:

```
c := linsolve( A, b );
```

$$c := \begin{bmatrix} 2 \\ -3 \\ 1 \end{bmatrix}$$

A matrix can also be defined by prescribing its *ij*th element as a function of the row index i and the column index i. For instance, the 3×3 Hilbert matrix is defined by

```
H3 := matrix( 3,3, (i,j) -> 1/(i+j-1) );
```

$$H3 := \begin{bmatrix} 1 & \frac{1}{2} & \frac{1}{3} \\ \frac{1}{2} & \frac{1}{3} & \frac{1}{4} \\ \frac{1}{3} & \frac{1}{4} & \frac{1}{5} \end{bmatrix}$$

The **`linalg`** package also includes an explicit Hilbert matrix function. For instance, the command

```
H4 := hilbert(4);
```

generates the 4×4 Hilbert matrix displayed in (2). Finally, the **`diag`**onal function can be used to generate the most ubiquitous of all special matrices — the identity matrices such as

```
I3 = diag(1,1,1);
```

$$\begin{bmatrix} 1 & 0 & 0 \\ 0 & 1 & 0 \\ 0 & 0 & 1 \end{bmatrix}$$

Using *Mathematica*

A particular matrix is entered in *Mathematica* as as list of row vectors, each row vector itself being a list of elements, as in the command

```
A = { {2,2,2}, {2,0,-2}, {1,-1,1} }
{{2, 2, 2}, {2, 0, -2}, {1, -1, 1}}
```

The **MatrixForm** function can be used to display **A** in standard matrix form:

```
A // MatrixForm
2    2    2
2    0    -2
1    -1   1
```

The inverse matrix $\mathbf{A}^{-1}$ is then calculated with the command

```
B = Inverse[A]
   1  1  1    1         1       1     1    1
{{-, -, -}, {-, 0, -(-)}, {-, -(-), -}}
   8  4  4    4         2       8     4    4
```

The period **.** denotes the matrix multiplication operator in *Mathematica*. Thus we can verify that **B** actually is the inverse matrix of **A** with the calculation

```
B . A // MatrixForm
1    0    0
0    1    0
0    0    1
```

Having defined the right-hand side constant (column) vector

```
b = { {0}, {2}, {6} }
{{0}, {2}, {6}}
```

in (1), we can then calculate the solution vector $\mathbf{c} = \mathbf{A}^{-1}\mathbf{b} = \mathbf{Bb}$ with the command

```
c = B . b
{{2}, {-3}, {1}}
```

For higher-dimensional linear systems the computation of the inverse matrix is not so efficient as immediate application of *Mathematica*'s **LinearSolve** function:

```
c = LinearSolve[ A, b ]
{{2}, {-3}, {1}}
```

The computation

```
A . c - b
{{0}, {0}, {0}}
```

then verifies that **c** is a solution of the equation **Ax** = **b**.

A matrix can also be defined by prescribing its *ij*th element as a function of the row index i and the column index i. For instance, the 3×3 Hilbert matrix is defined by

```
H3 = Table[ 1/(i+j-1), {i,1,3},{j,1,3} ]
      1  1     1  1  1    1  1  1
{{1, -, -}, {-, -, -}, {-, -, -}}
      2  3     2  3  4    3  4  5
```

```
H3 // MatrixForm
     1   1
     -   -
1    2   3
1    1   1
-    -   -
2    3   4
1    1   1
-    -   -
3    4   5
```

Finally, the **DiagonalMatrix** function can be used to generate the most ubiquitous of all special matrices — the identity matrices such as

```
I3 = DiagonalMatrix[ {1,1,1} ]
{{1, 0, 0}, {0, 1, 0}, {0, 0, 1}}
```

Actually, the identity matrix has its own function (of the dimension):

```
IdentityMatrix[3] // MatrixForm
1  0  0
0  1  0
0  0  1
```

Using MATLAB

A particular matrix can be entered either with a command of the form

```
A = [ [2 2 2]; [2 0 -2]; [1 -1 1] ]
A =
       2       2       2
       2       0      -2
       1      -1       1
```

where the row vectors of **A** are separated by semicolons, or with one of the simplest possible form

```
A = [ 2   2   2
      2   0  -2
      1  -1   1 ]
```

in which the matrix is "built" just as it looks The inverse matrix $\mathbf{A}^{-1}$ is then calculated with the command

```
B = inv(B)
B =
     0.1250      0.2500      0.2500
     0.2500           0     -0.5000
     0.1250     -0.2500      0.2500
```

The ordinary MATLAB multiplication operator `*` suffices to multiply matrices of appropriate dimensions. Thus we can verify that **B** actually is the inverse matrix of **A** with the calculation

```
B*A
ans =
       1       0       0
       0       1       0
       0       0       1
```

Having defined the right-hand side constant (column) vector

```
b = [0; 2; 6]
b =
       0
       2
       6
```

in (1), we can then calculate the solution vector $\mathbf{c} = \mathbf{A}^{-1}\mathbf{b} = \mathbf{Bb}$ with the command

```
c = B*b
```

```
c =
       2
      -3
       1
```

For higher-dimensional linear systems the computation of the inverse matrix is not so efficient as immediate application of MATLAB's "linear solve" function in the form of the operator \ that denotes "matrix *left* division":

```
c = A\b
c =
       2
      -3
       1
```

A matrix can also be defined with a double loop prescribing its *ij*th element as a function of the row index i and the column index j. For instance, the 3×3 Hilbert matrix $\mathbf{H}_3$ is defined by

```
for i = 1 : 3
    for j = 1 : 3
        H3(i,j) = 1/(i+j-1);
    end
end
```

using the notation $\mathbf{A}(i, j)$ for the *ij*th element of the matrix $\mathbf{A}$. After specifying "rational formatting" of output, we can display the result as

```
format rat
H3
H3 =
         1              1/2            1/3
       1/2              1/3            1/4
       1/3              1/4            1/5
```

MATLAB also includes an explicit Hilbert matrix function. For instance, the command

```
H4 = hilb(4)
```

generates the 4×4 Hilbert matrix $\mathbf{H}_4$ displayed in (2).

Finally, the **diag**onal function can be used to generate the most ubiquitous of all special matrices — the identity matrices such as

```
I4 = diag([1 1 1 1])
```

```
I4 =
     1     0     0     0
     0     1     0     0
     0     0     1     0
     0     0     0     1
```

Actually, the identity matrix has its own MATLAB function (of the dimension):

```
I5 = eye(5)
I5 =
     1     0     0     0     0
     0     1     0     0     0
     0     0     1     0     0
     0     0     0     1     0
     0     0     0     0     1
```

Application 5.2

Automatic Calculation of Eigenvalues and Eigenvectors

Most computational systems offer the capability to find eigenvalues and eigenvectors readily. For instance, for the matrix

$$\mathbf{A} = \begin{bmatrix} -0.5 & 0 & 0 \\ 0.5 & -0.25 & 0 \\ 0 & 0.25 & -0.2 \end{bmatrix} \quad (1)$$

of Example 2 in Section 5.2 of the text, the TI-86 commands

```
[[-0.5,0,0][0.5,-0.25,0][0,0,25,-0.2]] → A
eigVl A
                          {-.2, -.25, -.5}
eigVc A
                          [[0.00    0.00    1.00]
                           [0.00    1.00   -2.00]
                           [1.00   -5.00    1.67]]
```

produce the three eigenvalues -0.2, -0.25, and -0.5 of the matrix $\mathbf{A}$ and display beneath each its (column) eigenvector. Note that with results presented in decimal form, it is up to us to guess (and verify by matrix multiplication) that the exact eigenvector associated with the eigenvalue $\lambda = -\frac{1}{2}$ is $\mathbf{v} = \begin{bmatrix} 1 & -2 & \frac{5}{3} \end{bmatrix}^T$. The *Maple* commands

```
with(linalg):
A := matrix(3,3,[-0.5,0,0, 0.5,-0.25,0, 0,0.25,-0.2]);
eigenvects(A);
```

the *Mathematica* commands

```
A = {{-0.5,0,0}, {0.5,-0.25,0}, {0,0.25,-0.2}}
Eigensystem[A]
```

and the MATLAB commands

```
[-0.5,0,0; 0.5,-0.25,0; 0,0.25,-0.2]
[V, E] = eig(A)
```

(where $\mathbf{E}$ will be a diagonal matrix displaying the eigenvalues of $\mathbf{A}$ and the column vectors of $\mathbf{V}$ are the corresponding eigenvectors) produce similar results. You can use these commands to find the eigenvalues and eigenvectors needed for any of the problems in Section 5.2 of the text.

Brine Tank Investigations

Consider a linear cascade of 5 full brine tanks whose volumes $v_1,\ v_2,\ v_3,\ v_4,\ v_5$ are given by

$$v_i \;=\; 10\,d_i \qquad\qquad \text{(gallons)}$$

where $d_1,\ d_2,\ d_3,\ d_4,\ d_5$ are the first five *distinct* non-zero digits of ***your*** student ID number. (Pick additional digits at random if your ID number has less than five distinct non-zero digits.)

Initially, Tank 1 contains one pound of salt per gallon of brine, whereas the remaining tanks contain pure water. The brine in each tank is kept thoroughly mixed, and the flow rate out of each tank is $r_i = 10$ gal/min. Your task is to investigate the subsequent amounts $x_1(t),\ x_2(t),\ x_3(t),\ x_4(t),\ x_5(t)$ of salt (in pounds) present in these brine tanks.

Open System

If fresh water flows into Tank 1 at the rate of 10 gal/min, then these functions satisfy the system

$$\begin{aligned} x_1' &= \qquad\;\; -k_1x_1 \\ x_2' &= +k_1x_1 - k_2x_2 \\ x_3' &= +k_2x_2 - k_3x_3 \\ x_4' &= +k_3x_3 - k_4x_4 \\ x_5' &= +k_4x_4 - k_5x_5 \end{aligned} \tag{2}$$

where $k_i = r_i/v_i$ for i = 1, 2, ..., 5. Find the eigenvalues $\lambda_1, \lambda_2, \lambda_3, \lambda_4, \lambda_5$ and the corresponding eigenvectors $\mathbf{v}_1, \mathbf{v}_2, \mathbf{v}_3, \mathbf{v}_4, \mathbf{v}_5$ of the this system's coefficient matrix in order to write the general solution in the form

$$\mathbf{x}(t) \;=\; c_1\mathbf{v}_1e^{\lambda_1 t} + c_2\mathbf{v}_2e^{\lambda_2 t} + c_3\mathbf{v}_3e^{\lambda_3 t} + c_4\mathbf{v}_4e^{\lambda_4 t} + c_5\mathbf{v}_5e^{\lambda_5 t}\,. \tag{3}$$

Use the given initial conditions to find the values of the constants c_1, c_2, c_3, c_4, c_5. Then observe that each $x_i(t) \to 0$ as $t \to \infty$, and explain why you would anticipate this result. Plot the graphs of the component functions $x_1(t), x_2(t), x_3(t), x_4(t), x_5(t)$ of $\mathbf{x}(t)$ on a single picture, and finally note (at least as close as the mouse will take you) the maximum amount of salt that is ever present in each tank.

Closed System

If Tank 1 receives as inflow (rather than fresh water) the outflow from Tank 5, then the first equation in (2) is replaced with the equation

$$x_1' = +k_5 x_5 - k_1 x_1. \qquad (4)$$

Assuming the same initial conditions as before, find the explicit solution of the form in (3). Now show that — in this *closed* system of brine tanks — as $t \to \infty$ the salt originally in Tank 1 distributes itself with uniform concentration throughout the various tanks. A plot should make this point rather vividly.

Maple, *Mathematica*, and MATLAB techniques that will be useful for these brine tank investigations are illustrated in the sections that follow. We consider the "open system" of three brine tanks that is shown in Fig. 5.2.2 of the text (see Example 2 of Section 5.2). The vector $\mathbf{x}(t) = [x_1(t)\ \ x_2(t)\ \ x_3(t)]^T$ of salt amounts (in the three tanks) satisfies the linear system

$$\frac{d\mathbf{x}}{dt} = \mathbf{Ax} \qquad (5)$$

where $\mathbf{A}$ is the 3×3 matrix in (1). If initially Tank 1 contains 15 pounds of salt and the other two tanks contain pure water, then the initial vector is $\mathbf{x}(0) = \mathbf{x}_0 = [15\ \ 0\ \ 0]^T$.

Using *Maple*

We begin by entering (as indicated previously) the coefficient matrix in (1),

```
with(linalg):
A := matrix(3,3,[-0.5,0,0, 0.5,-0.25,0, 0,0.25,-0.2]);
```

$$A := \begin{bmatrix} -0.5 & 0 & 0 \\ 0.5 & -0.25 & 0 \\ 0 & 0.25 & -0.2 \end{bmatrix}$$

and the initial vector

```
x0 := matrix(3,1, [15,0,0]);
```

$$x0 := \begin{bmatrix} 15 \\ 0 \\ 0 \end{bmatrix}$$

The eigenvalues and eigenvectors of **A** are calculated with the command

```
eigs := eigenvects(A);
```

$$eigs := [-.5, 1, \{[1\ \ -2.000000000\ \ 1.666666667]\}],$$
$$[-.25, 1, \{[0\ \ 1\ \ -5.000000000]\}],$$
$$[-.2, 1, \{[0\ \ 0\ \ 1]\}]$$

Thus the first eigenvalue λ_1 and its associated eigenvector $\mathbf{v}_1$ are given by

```
eigs[1][1];
```

$$-.5$$

```
eigs[1][3][1];
```

$$[1\quad -2.000000000\quad 1.666666667]$$

We therefore record the three eigenvalues

```
L1 := eigs[1][1]:
L2 := eigs[2][1]:
L3 := eigs[3][1]:
```

and the corresponding three eigenvectors

```
v1 := matrix(1,3, eigs[1][3][1]):
v2 := matrix(1,3, eigs[2][3][1]):
v3 := matrix(1,3, eigs[3][3][1]):
```

The matrix **V** with these three column vectors is then defined by

```
V := transpose(stackmatrix(v1,v2,v3));
```

$$V := \begin{bmatrix} 1 & 0 & 0 \\ -2.000000000 & 1 & 0 \\ 1.666666667 & -5.000000000 & 1 \end{bmatrix}$$

To find the constants $c_1,\ c_2,\ c_3$ in the solution

$$\mathbf{x}(t) = c_1\mathbf{v}_1 e^{\lambda_1 t} + c_2\mathbf{v}_2 e^{\lambda_2 t} + c_3\mathbf{v}_3 e^{\lambda_3 t} \tag{6}$$

we need only solve the system $\mathbf{Vc} = \mathbf{x_0}$:

```
c := linsolve(V,x0);
```

$$c := \begin{bmatrix} 15. \\ 30. \\ 125.0000000 \end{bmatrix}$$

Recording the values of these three constants,

```
c1  := c[1,1]:
c2  := c[2,1]:
c3  := c[3,1]:
```

we can finally calculate the solution

```
x := evalm(c1*v1*exp(L1*t) + c2*v2*exp(L2*t) +
                c3*v3*exp(L3*t)):
```

in (6) with component functions

```
x1  := x[1,1];

x2  := x[1,2];

x3  := x[1,3];
```

$$x1 := 15.\, e^{-.5t}$$

$$x2 := -30.00000000\, e^{-.5t} + 30.\, e^{-.25t}$$

$$x2 := 25.00000001\, e^{-.5t} - 150.00000000\, e^{-.25t} + 125.00000000\, e^{-.2t}$$

The command

```
plot( {x1,x2,x3}, t = 0..30 );
```

produces the figure on the next page showing the graphs of the functions $x_1(t)$, $x_2(t)$, and $x_3(t)$ giving the amounts of salt in the three tanks. We can approximate the maximum value of each $x_i(t)$ by mouse-clicking on the apex of the appropriate graph.

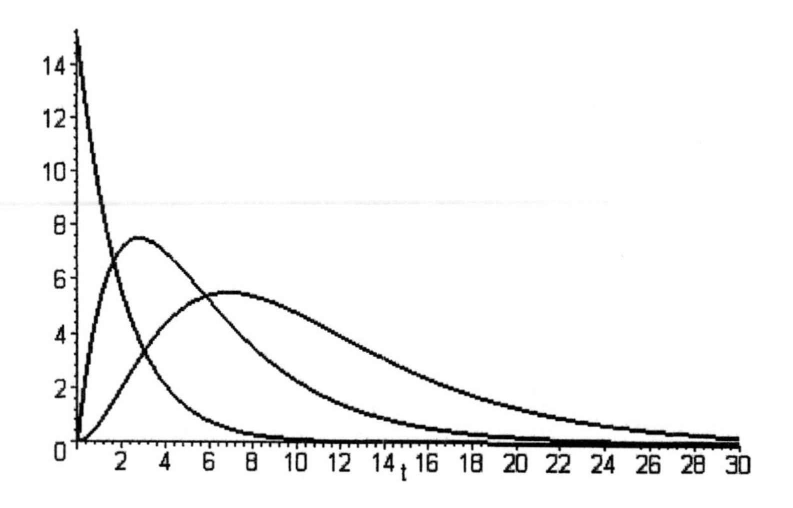

Using *Mathematica*

We begin by entering (as indicated previously) the coefficient matrix in (1),

```
A = {{-0.50 ,0,0}, {0.50,-0.25,0}, {0,0.25,-0.20}};
A // MatrixForm
```

```
-0.5     0        0
0.5      -0.25    0
0        0.25     -0.2
```

and the initial vector

```
x0 = {{15}, {0}, {0}};  x0 // MatrixForm
```

```
15
0
0
```

The eigenvalues and eigenvectors of **A** are calculated with the command

```
eigs = Eigensystem[A]
```

```
{{-0.5,-0.25,-0.2}, {{0.358569, -0.717137, 0.597614},
   {0., 0.196116, -0.980581}, {0., 0., 1.}}}
```

Thus the first eigenvalue λ_1 and its associated eigenvector $\mathbf{v}_1$ are given by

```
eigs[[1,1]]
```

```
-0.5
```

```
eigs[[2,1]]
```

```
{0.358569, -0.717137, 0.597614}
```

We therefore record the three eigenvalues

```
L1 = eigs[[1,1]];
L2 = eigs[[1,2]];
L3 = eigs[[1,3]];
```

and the corresponding three eigenvectors

```
v1 = eigs[[2,1]];
v2 = eigs[[2,2]];
v3 = eigs[[2,3]];
```

The matrix **V** having these three eigenvectors as its *column* vectors is then defined by

```
V = Transpose[ {v1, v2, v3} ];
V // MatrixForm
```

```
0.358569     0.           0.
-0.717137    0.196116     0.
0.597614     -0.980581    1.
```

To find the constants c_1, c_2, c_3 in the solution

$$\mathbf{x}(t) = c_1\mathbf{v}_1 e^{\lambda_1 t} + c_2\mathbf{v}_2 e^{\lambda_2 t} + c_3\mathbf{v}_3 e^{\lambda_3 t} \tag{6}$$

we need only solve the system $\mathbf{Vc} = \mathbf{x_0}$:

```
c = LinearSolve[V,x0]
```

```
{{41.833}, {152.971}, {125.}}
```

Recording the values of these three constants,

```
c1 = c[[1,1]];
c2 = c[[2,1]];
c3 = c[[3,1]];
```

we can finally calculate the solution

```
x = c1*v1*Exp[L1*t]+c2*v2*Exp[L2*t]+c3*v3*Exp[L3*t];
```

in (6) with component functions

```
x1 = x[[1]]
x2 = x[[2]]
x3 = x[[3]]
```

$$\frac{0.}{e^{0.25t}}+\frac{15.}{e^{0.5t}}+\frac{0.}{e^{0.2t}}$$

$$\frac{30.}{e^{0.25t}}-\frac{30.}{e^{0.5t}}+\frac{0.}{e^{0.2t}}$$

$$-\frac{150.}{e^{0.25t}}+\frac{25.}{e^{0.5t}}+\frac{125.}{e^{0.2t}}$$

The command

```
Plot[ {x1,x2,x3}, {t,0,30} ]
```

then produces the figure below, showing the graphs of the functions $x_1(t)$, $x_2(t)$, and $x_3(t)$ that give the amounts of salt in the three tanks. We can approximate the maximum value of each $x_i(t)$ by mouse-clicking on the apex of the appropriate graph.

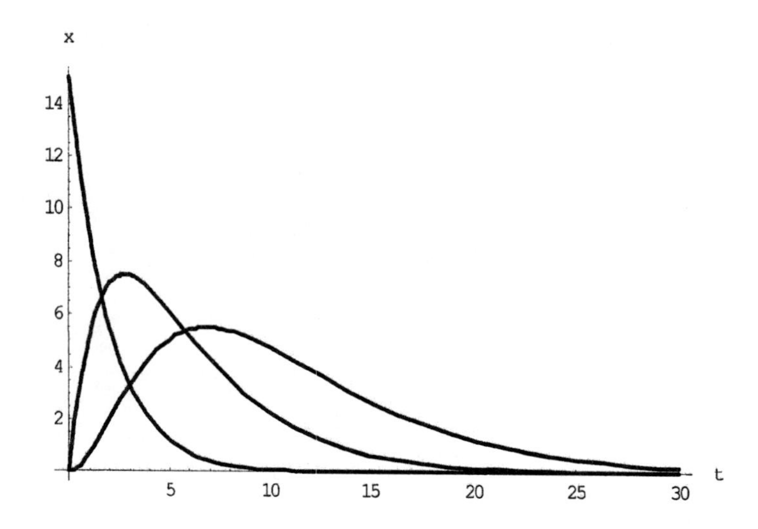

Using MATLAB

We begin by entering (as indicated previously) the coefficient matrix in (1),

```
A = [-0.5      0      0
      0.5  -0.25      0
        0   0.25   -0.2 ]
A =
   -0.5000         0         0
    0.5000   -0.2500         0
         0    0.2500   -0.2000
```

and the initial vector

```
x0 = [15; 0; 0];
```

The eigenvalues and eigenvectors of **A** are calculated with the command

```
[V, E] = eig(A)
V =
         0         0    0.3586
         0    0.1961   -0.7171
    1.0000   -0.9806    0.5976

E =
   -0.2000         0         0
         0   -0.2500         0
         0         0   -0.5000
```

The eigenvalues of **A** are the diagonal elements

```
L = diag(E)
L =
   -0.2000
   -0.2500
   -0.5000

L1 = L(1);      L2 = L(2);      L3 = L(3);
```

of the matrix **E**. The associated eigenvectors are the corresponding column vectors

```
v1 = V(:,1);   v2 = V(:,2);   v3 = V(:,3);
```

of the matrix **V**. To find the constants c_1, c_2, c_3 in the solution

$$\mathbf{x}(t) = c_1\mathbf{v}_1 e^{\lambda_1 t} + c_2\mathbf{v}_2 e^{\lambda_2 t} + c_3\mathbf{v}_3 e^{\lambda_3 t} \qquad (6)$$

we need only solve the system $\mathbf{Vc} = \mathbf{x}_0$:

```
c = V\x0;    c'
ans =
  125.0000  152.9706   41.8330
```

Recording the values of these three constants,

```
c1 = c(1);    c2 = c(2);    c3 = c(3);
```

and defining an appropriate range

```
t = 0 : 0.1 : 30;
```

of values of t, we can finally calculate the solution

```
x = c1*v1*exp(L1*t)+c2*v2*exp(L2*t)+c3*v3*exp(L3*t);
```

in (6). We plot its three component functions

```
x1 = x(1,:);    x2 = x(2,:);    x3 = x(3,:);
```

using the command

```
plot( t,x1, t,x2, t,x3 )
```

The resulting figure (just like those exhibited in the preceding *Maple* and *Mathematica* discussions) shows the graphs of the functions $x_1(t)$, $x_2(t)$, and $x_3(t)$ giving the amounts of salt in the three tanks. We can approximate the maximum value of each $x_i(t)$ by mouse-clicking (after **ginput**) on the apex of the appropriate graph.

Complex Eigenvalues

Finally we consider the closed system of three brine tanks that is shown in Fig. 5.2.5 of the text (see Example 4 of Section 5.2). The vector $\mathbf{x}(t)=[x_1(t)\ \ x_2(t)\ \ x_3(t)]^T$ of salt amounts (in the three tanks) satisfies the linear system

$$\frac{d\mathbf{x}}{dt} = A\mathbf{x} \tag{5}$$

where $\mathbf{A}$ now is the 3×3 matrix defined by

```
A = [-0.2    0    0.2
      0.2  -0.4    0
       0    0.4  -0.2 ]
A =
   -0.2000         0    0.2000
    0.2000   -0.4000         0
         0    0.4000   -0.2000
```

If initially Tank 1 contains 10 pounds of salt and the other two tanks contain pure water, then the initial vector $\mathbf{x}(0)=\mathbf{x}_0$ is defined by

```
x0 = [10; 0; 0];
```

The eigenvalues and eigenvectors of $\mathbf{A}$ are calculated using the command

```
[V, E] = eig(A)
V =
   -0.6667       0.4976 + 0.0486i       0.4976 - 0.0486i
   -0.3333       0.0486 - 0.4976i       0.0486 + 0.4976i
   -0.6667      -0.5462 + 0.4491i      -0.5462 - 0.4491i

E =
   -0.0000       0                      0
         0      -0.4000 + 0.2000i       0
         0                    0        -0.4000 - 0.2000i
```

Now we see complex conjugate pairs of eigenvalues and eigenvectors, but let us nevertheless proceed without fear, hoping that "ordinary" real-valued solution functions will somehow result. First we pick off and record the eigenvalues that appear as the diagonal elements of the matrix **E**.

```
L = diag(E);
L'

ans =
   -0.0000      -0.4000 - 0.2000i      -0.4000 + 0.2000i

L1 = L(1);
L2 = L(2);
L3 = L(3);
```

The associated eigenvectors are the corresponding column vectors

```
v1 = V(:,1);
v2 = V(:,2);
v3 = V(:,3);
```

of the matrix **V**. To find the constants $c_1,\ c_2,\ c_3$ in the solution

$$\mathbf{x}(t) = c_1\mathbf{v}_1e^{\lambda_1 t} + c_2\mathbf{v}_2e^{\lambda_2 t} + c_3\mathbf{v}_3e^{\lambda_3 t} \tag{6}$$

we need only solve the system $\mathbf{Vc} = \mathbf{x_0}$:

```
c = V\x0;
c'
ans =
   -6.0000+0.0000i   5.7773+2.5735i   5.7773-2.5735i
```

Recording the values of these three constants,

```
c1 = c(1);
c2 = c(2);
c3 = c(3);
```

and defining an appropriate range

```
t = 0 : 0.1 : 30;
```

of values of t, we can finally calculate the solution

```
x = c1*v1*exp(L1*t)+c2*v2*exp(L2*t)+c3*v3*exp(L3*t);
```

We can plot the three component functions

```
x1 = x(1,:);
x2 = x(2,:);
x3 = x(3,:);
```

using the command

```
plot( t,x1, t,x2, t,x3 )
```

The result is the figure below. It shows the graphs of the functions $x_1(t)$, $x_2(t)$, and $x_3(t)$ that give the amounts of salt in the three tanks. Is it clear to you that the three solution curves "level off" as $t \to \infty$ in a way that exhibits a long-term uniform concentration of salt throughout the system?

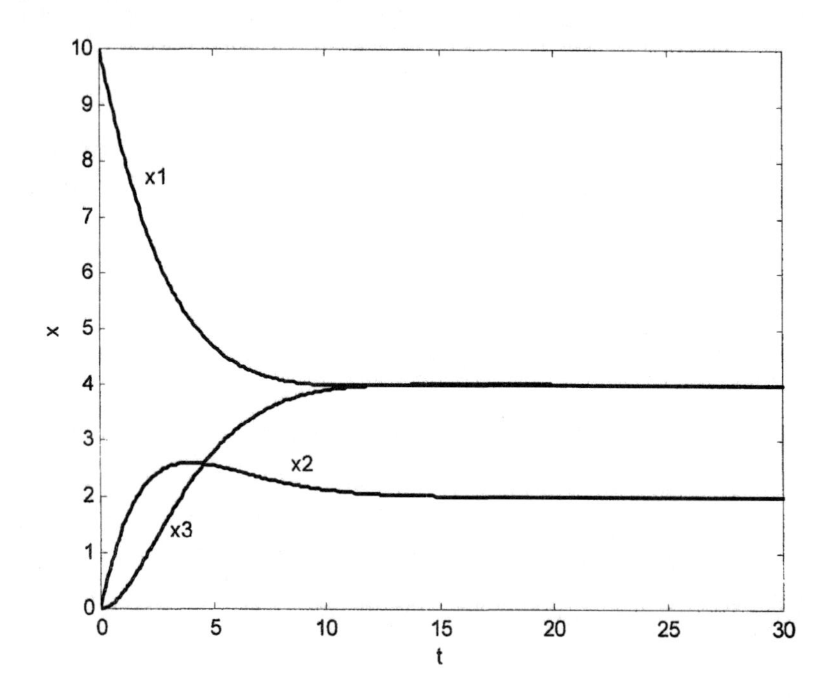

Further Investigation

If our system consists of only two equations in the two unknowns x_1 and x_2, then we can describe the behavior of the solutions not only by plotting them individually as functions of t, as we have done above, but also by drawing the *phase portrait* of the system. This topic will be covered in more detail in Chapter 6, but here we want to apply this idea in conjunction with the eigenvalue method to gain some preliminary insights into the behavior of linear homogeneous constant-coefficient systems.

We are going to study the following system, whose coefficients depend upon a constant k:

$$x_1' = (k+2)x_1 + 2x_2$$
$$x_2' = x_1 + (k+1)x_2$$

We can write this system in the form $\frac{d\mathbf{x}}{dt} = \mathbf{Ax}$, where $\mathbf{A}$ denotes the 2×2 matrix

$\begin{bmatrix} k+2 & 2 \\ 1 & k+1 \end{bmatrix}$ and $\mathbf{x}(t)$ denotes the column vector $\begin{bmatrix} x_1(t) \\ x_2(t) \end{bmatrix}$.

We illustrate this investigation using *Mathematica*; *Maple* and MATLAB users can substitute equivalent commands. First we set up the system in *Mathematica*...

```
A := {{k + 2, 2}, {1, k + 1}}
x := {x1[t], x2[t]}
col1 = A[[1]].x
col2 = A[[2]].x
```

...and then find symbolically the solution curves through all points (i, j) in the plane, where i and j take on all integer values from -2 to 2:

```
temp = Table[DSolve[{x1'[t] -> col1, x2'[t] -> col2,
                x1[0] -> i, x2[0] -> j}, x, t],
                {i, -2, 2}, {j, -2, 2}];
partsolns = Flatten[temp, 2]
```

A graph of these solution curves together is called a phase portrait of the system. Now we can make an animation showing the phase portrait as k varies from -2 to 2 in increments of 0.1:

```
Table[ParametricPlot[
        Evaluate[{x1[t], x2[t]} /.partsolns],
        {t, -50, 50}, PlotRange -> {{-2, 2}, {-2, 2}},
        AspectRatio -> Automatic], {k, -2, 2, 0.1}];
```

Double-click on any of the frames to play the animation (during play you can use the number keys 1,2,3,... to adjust the speed of play), and then answer the following questions:

- What do all of the frames of the animation seem to have in common? Can you translate this into something that's true of the matrix **A** regardless of the value of k?
- What becomes of all of the solutions you found symbolically when $k = 0$? Which frame of the animation reflects this? (The *Mathematica* command `partsolns/.k -> 0` might help.)
- What about the matrix **A** *does* change as k varies? (Yes, the coefficients, but more to the point...)

Finally, use *Mathematica* as above to find the eigenvalues and eigenvectors of **A**; were your preceding answers borne out?

Application 5.3

Earthquake-Induced Vibrations of Multistory Buildings

In this application you are to investigate the response to transverse earthquake ground oscillations of a seven-story building like the one illustrated in Fig. 5.3.15 of the text. Suppose that each of the seven [aboveground] floors weighs 16 tons, so the mass of each is $m = 1000$ (slugs). Also assume a horizontal restoring force of $k = 5$ (tons per foot) between adjacent floors. That is, the internal forces in response to horizontal displacements of the individual floors are those shown in Fig. 5.3.16. It follows that the free transverse oscillations indicated in Fig. 5.3.15 satisfy the equation $\mathbf{Mx}'' = \mathbf{Kx}$ with $n = 7$ and $m_i = 1000$, $k_i = 10{,}000$ (lb/ft) for $1 \le i \le 7$. The system then reduces to the form $\mathbf{x}'' = \mathbf{Ax}$ with

$$\mathbf{A} = \begin{bmatrix} -20 & 10 & 0 & 0 & 0 & 0 & 0 \\ 10 & -20 & 10 & 0 & 0 & 0 & 0 \\ 0 & 10 & -20 & 10 & 0 & 0 & 0 \\ 0 & 0 & 10 & -20 & 10 & 0 & 0 \\ 0 & 0 & 0 & 10 & -20 & 10 & 0 \\ 0 & 0 & 0 & 0 & 10 & -20 & 10 \\ 0 & 0 & 0 & 0 & 0 & 10 & -10 \end{bmatrix} \qquad (1)$$

Once the matrix **A** has been entered, the TI-86 command **`eigVl A`** takes only about 15 seconds to calculate the seven eigenvalues shown in the λ-row of the table below. Alternatively, you can use the *Maple* **`with(linalg)`** command **`eigenvals(A)`**, the *Mathematica* command **`Eigenvalues[A]`**, or the MATLAB command **`eig(A`**. Then the remaining entries $\omega = \sqrt{-\lambda}$ and $P = 2\pi / \omega$ showing the natural frequencies and periods of oscillation of the seven-story building are readily calculated. Note that a typical earthquake producing ground oscillations with a period of 2 seconds is uncomfortably close to the fifth natural period 1.9869 seconds of the building.

i	1	2	3	4	5	6	7
λ	–38.2709	-33.3826	–26.1803	–17.9094	10.0000	–3.8197	–0.4370
ω	6.1863	5.7778	5.1167	4.2320	3.1623	1.9544	0.6611
P (sec)	1.1057	1.0875	1.2280	1.4847	1.9869	3.2149	9.5042

A horizontal earthquake oscillation $E \cos \omega t$ of the ground, with amplitude E and acceleration $a = -E\omega^2 \cos \omega t$, produces an opposite inertial force $F = ma = mE\omega^2 \cos \omega t$ on each floor of the building. The resulting nonhomogeneous linear system is

$$\mathbf{x}'' = \mathbf{A}\mathbf{x} + (E\omega^2 \cos \omega t)\,\mathbf{b} \tag{2}$$

where $\mathbf{b} = [1 \;\; 1 \;\; 1 \;\; 1 \;\; 1 \;\; 1 \;\; 1]^T$ and $\mathbf{A}$ is the matrix of Eq. (1). Figure 5.3.18 in the text shows a plot of *maximal amplitude* (for the forced oscillations of any single floor) versus the *period* of the earthquake vibrations. The spikes correspond to the first six of the seven resonance frequencies. We see, for instance, that while an earthquake with period 2 seconds likely would produce destructive resonance vibrations in the building, it probably would be unharmed by an earthquake with period 2.5 seconds. Different buildings have different natural frequencies of vibration, and so a given earthquake may demolish one building, but leave untouched the one next door. This type of apparent anomaly was observed in Mexico City after the devastating earthquake of September 19, 1985.

Investigation

For your personal seven-story building to investigate, let the weight (in tons) of each story equal the largest digit of your student ID number and let k (in tons/ft) equal the smallest nonzero digit. Produce numerical and graphical results like those illustrated in Figs. 5.3.16 and 5.3.18 of the text. Is your building susceptible to likely damage from an earthquake with period in the 2 to 3 second range?

You might like to begin by working manually the following warm-up problems.

1. Find the periods of the natural vibrations of a building with two aboveground floors, with each weighing 16 tons and with each restoring force being $k = 5$ tons/ft.

2. Find the periods of the natural vibrations of a building with three aboveground floors, with each weighing 16 tons and with each restoring force being $k = 5$ tons/ft.

3. Find the natural frequencies and natural modes of vibration of a building with three aboveground floors as in Problem 2, except that the upper two floors weigh 8 tons instead of 16 tons. Give the ratios of the amplitudes A, B, and C of the oscillations of the three floors in the form $A : B : C$ with $A = 1$.

4. Suppose that the building of Problem 3 is subject to an earthquake in which the ground undergoes horizontal sinusoidal oscillations with a period of 3 seconds and an amplitude of 3 inches. Find the amplitudes of the resulting steady periodic oscillations of the three above-ground floors. Assume the fact that a motion $E \cos \omega t$ of the ground, with amplitude E and acceleration $a = -E\omega^2 \cos \omega t$, produces an opposite inertial force $F = ma = mE\omega^2 \cos \omega t$ on a floor of mass m.

A Three-Mass Automobile Model

In the sections that follow, we illustrate appropriate *Maple*, *Mathematica*, and MATLAB techniques by analyzing the natural frequencies of vibration of a car that is modeled by a system of three masses and four springs. Suppose that

- mass m_1 is connected to the chassis by spring k_1;
- masses m_1 and m_2 are connected by spring k_2;
- masses m_2 and m_3 are connected by spring k_3; and
- masses m_1 and m_3 are connected by spring k_4.

The corresponding linear system $\mathbf{Mx}'' = \mathbf{Kx}$ has coefficient matrices

$$\mathbf{M} = \begin{bmatrix} m_1 & 0 & 0 \\ 0 & m_2 & 0 \\ 0 & 0 & m_3 \end{bmatrix} \tag{3}$$

and

$$\mathbf{K} = \begin{bmatrix} -(k_1+k_2+k_4) & k_2 & k_4 \\ k_2 & -(k_2+k_3) & k_3 \\ k_4 & k_3 & -(k_3+k_4) \end{bmatrix}. \tag{4}$$

The displacement vector $\mathbf{x}(t) = \begin{bmatrix} x_1(t) & x_2(t) & x_3(t) \end{bmatrix}^T$ of the three-mass system then satisfies the equation

$$\mathbf{x}'' = \mathbf{Ax} \tag{5}$$

with coefficient matrix $\mathbf{A} = \mathbf{M}^{-1}\mathbf{K}$.

We will use the numerical values $m_1 = 40$, $m_2 = 20$, $m_3 = 40$ (in slugs) and $k_1 = 5000$, $k_2 = 1000$, $k_3 = 2000$, $k_4 = 3000$ (in lbs/ft). We want to find the three natural frequencies $\omega_1, \omega_2, \omega_3$ of oscillation of this three-mass system modeling our car. If the car is driven with velocity v (ft/sec) over a washboard surface shaped like a cosine curve with a wavelength of $a = 30$ feet, then the result is a periodic force on the car with frequency $\omega = 2\pi v / a$. We would expect the car to experience resonance vibrations when this forcing frequency equals one of the car's natural frequencies.

Using *Maple*

First we define the masses

```
m1 := 40:    m2 := 20:    m3 := 40:
```

and the spring constants

```
k1 := 4000:    k2 := 1000:    k3 := 2000:    k4 := 3000:
```

Then the mass, stiffness, and coefficient matrices of our system are defined by

```
with(linalg):
M := diag(m1,m2,m3);
```

$$M := \begin{bmatrix} 40 & 0 & 0 \\ 0 & 20 & 0 \\ 0 & 0 & 40 \end{bmatrix}$$

```
K := matrix(3,3, [-(k1+k2+k4),      k2,          k4,
                        k2,      -(k2+k3),      k3,
                        k4,         k3,     -(k3+k4)] );
```

$$K := \begin{bmatrix} -8000 & 1000 & 3000 \\ 1000 & --3000 & 2000 \\ 3000 & 2000 & -5000 \end{bmatrix}$$

```
A := evalm( inverse(M) &* K );
```

$$A := \begin{bmatrix} -200 & 25 & 75 \\ 50 & -150 & 100 \\ 75 & 50 & -125 \end{bmatrix}$$

The eigenvalues of **A** are given by

```
eigs := evalf(eigenvals(A));
```

$$eigs := -200., -27.8129452, -247.1870548$$

and we record them in increasing order of magnitude:

```
L := matrix(1,3, [eigs[2],eigs[1],eigs[3]] );
```

$$L := \begin{bmatrix} -27.8129452 & -200. & -247.1870548 \end{bmatrix}$$

The corresponding natural frequencies of the system are then given by

```
w := matrix(1,3, [sqrt(-L[1,1]), sqrt(-L[1,2]),
                   sqrt(-L[1,3])] );
```

$$w := [5.273797986 \quad 14.14213562 \quad 15.72218353]$$

When the car is driven over a washboard surface with wavelength

```
a := 30:        # in feet
```

the resulting critical velocities are given (in ft/s) by

```
v := evalf(evalm( (w*a)/(2*Pi) )):
```

and — since 88 ft/sec corresponds to 60 miles per hour — by

```
mph := evalm( 60*v/88 );
```

$$mph := [17.16854355 \quad 46.03890251 \quad 51.18265687]$$

in miles per hour. If the car is accelerated from 0 to 60 mph, it would therefore experience resonance vibrations as it passes through the speeds of 17, 46, and 51 miles per hour.

Using *Mathematica*

First we define the masses

```
m1 = 40;    m2 = 20;    m3 = 40;
```

and the spring constants

```
k1 = 4000;    k2 = 1000;    k3 = 2000;    k4 = 3000;
```

Then the mass, stiffness, and coefficient matrices of our system are defined by

```
M = DiagonalMatrix[{m1,m2,m3}];
M // MatrixForm

40   0    0
0    20   0
0    0    40
```

```
K = { {-(k1+k2+k4),    k2,          k4    },
      {      k2,     -(k2+k3),      k3    },
      {      k4,        k3,     -(k3+k4)} };
K // MatrixForm

-8000    1000     3000
1000     -3000    2000
3000     2000     -5000
```

```
A = Inverse[M] . K;
A // MatrixForm
```

```
-200    25      75
50      -150    100
75      50      -125
```

The eigenvalues of **A** are given by

```
eigs = Eigenvalues[A] // N
```

```
{-200., -247.187, -27.8129}
```

and we sort them in increasing order of magnitude:

```
L = Reverse[Sort[eigs]]
```

```
{-27.8129, -200., -247.187}
```

The corresponding natural frequencies of the system are then given by

```
w = Sqrt[-L]
```

```
{5.2738, 14.1421, 15.7222}
```

When the car is driven over a washboard surface with wavelength

```
a = 30;          (* in feet  *)
```

the resulting critical velocities are given (in ft/s) by

```
v = w*a/(2*Pi) // N;
```

and — since 88 ft/sec corresponds to 60 miles per hour — by

```
mph = 60*v/88
```

```
{17.1685, 46.0389, 51.1827}
```

in miles per hour. If the car is accelerated from 0 to 60 mph, it would therefore experience resonance vibrations as it passes through the speeds of 17, 46, and 51 miles per hour.

Using MATLAB

First we define the masses

```
m1 = 40;    m2 = 20;    m3 = 40;
```

and the spring constants

```
k1 = 4000;    k2 = 1000;    k3 = 2000;    k4 = 3000;
```

Then the mass, stiffness, and coefficient matrices of our system are defined by

```
M = diag([m1  m2  m3])
M =
    40     0     0
     0    20     0
     0     0    40

K = [-(k1+k2+k4),        k2,           k4;
           k2,       -(k2+k3),         k3;
           k4,           k3,       -(k3+k4) ]
K =
       -8000          1000          3000
        1000         -3000          2000
        3000          2000         -5000

A = M\K
A =
  -200     25     75
    50   -150    100
    75     50   -125
```

The eigenvalues of **A** are given by

```
eigs = eig(A)'
eigs =
  -27.8129 -247.1871 -200.0000
```

and we sort them in increasing order of magnitude:

```
L = fliplr(sort(eigs))
L =
  -27.8129 -200.0000 -247.1871
```

The corresponding natural frequencies of the system are then given by

```
w = sqrt(-L)
w =
    5.2738   14.1421   15.7222
```

When the car is driven over a washboard surface with wavelength

```
a = 30;      % in feet
```

the resulting critical velocities are given by

```
v = w*a/(2*pi);
```

and — since 88 ft/sec corresponds to 60 miles per hour — by

```
mph = 60*v/88
mph =
   17.1685   46.0389   51.1827
```

in miles per hour. If the car is accelerated from 0 to 60 mph, it would therefore experience resonance vibrations as it passes through the speeds of 17, 46, and 51 miles per hour.

Resonance Vibrations of the 7-Story Building

We describe here how MATLAB was used to generate Figure 5.3.18 in the text, showing maximal amplitude of oscillations (for any single floor) as a function the period P of the earthquake. First, the commands

```
V = ones(1,6);
A = 10*diag(V,1) - 20*eye(7) + 10*diag(V,-1);
A(7,7) = -10
```

were entered to set up the coefficent matrix $\mathbf{A}$ of Eq. (1). If we substitute the trial solution

$$\mathbf{x} = \mathbf{v} \cos \omega t$$

(with undetermined coefficient vector $\mathbf{v}$) in Eq. (2), we get the matrix equation

$$(A + \omega^2 I)\mathbf{v} = -E\omega^2 \mathbf{b}$$

that is readily solved numerically for the amplitude vector $\mathbf{v}$ of the resulting forced vibrations of the individual floors of the building. The following MATLAB function **amp** does this and then selects the maximal amplitude of forced vibration of any single floor of the building in response to an earthquake vibration with period P.

```
function  y = amp(A,P)
E = 0.25;                % earthquake amplitude
n = size(A);
n = n(1,1);              % dimension of system
Id = eye(n);             % n by n identity matrix
b = ones(n,1);           % constant vector
k = size(P);
```

```
k = k(1,2);              % length of input vector P
y = ones(1,k);           % initialize y
for j = 1:k
    w = 2*pi/P(j);
    V = (A + w*w*Id)\(-E*w*w*b);      % solution of
    y(j) = max( abs(V) );             % linear system
    end
```

To calculate the maximal response **y** to vibrations with periods 0.01, 0.02, 0.03, ..., 4.99, 5.00 we need only define the vector **P** of periods and invoke the function **amp**.

```
P = 0.01:0.01:5;
y = amp(A,P);
```

A modern microcomputer solves the 500 necessary 7-by-7 linear systems in a matter of seconds. Finally, we need only

```
plot(P, y)
```

to see our results on the screen.

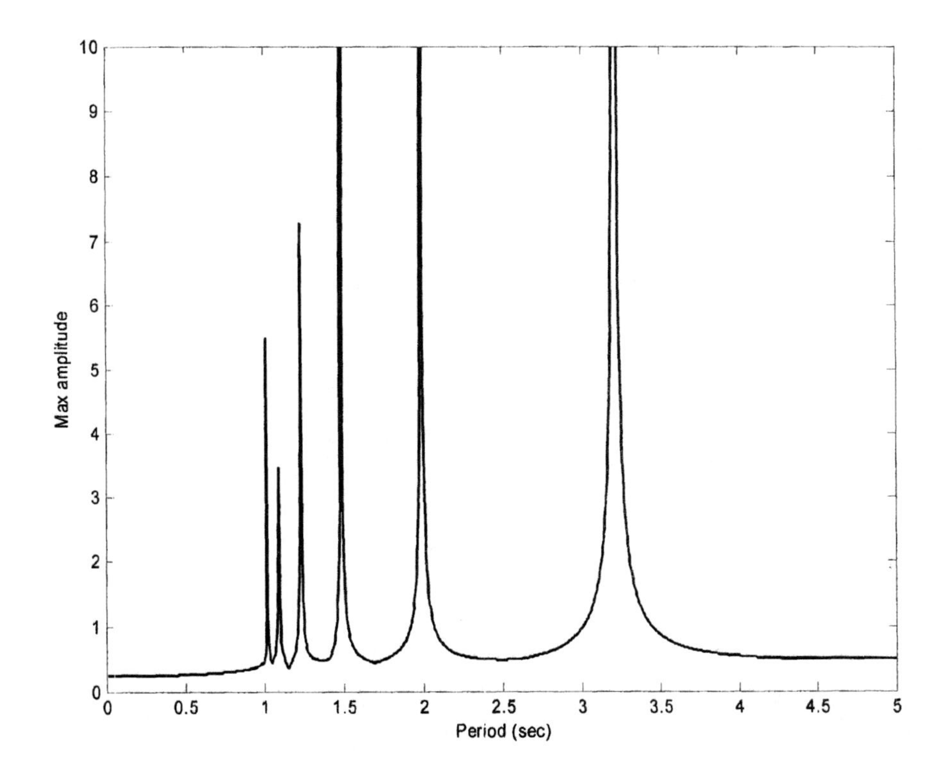

Application 5.4
Defective Eigenvalues and Generalized Eigenvectors

The goal of this application is the solution of the linear systems like

$$\mathbf{x}' = \mathbf{A}\mathbf{x}, \tag{1}$$

where the coefficient matrix is the exotic 5-by-5 matrix

$$\mathbf{A} = \begin{bmatrix} -9 & 11 & -21 & 63 & -252 \\ 70 & -69 & 141 & -421 & 1684 \\ -575 & 575 & -1149 & 3451 & -13801 \\ 3891 & -3891 & 7782 & -23345 & 93365 \\ 1024 & -1024 & 2048 & -6144 & 24572 \end{bmatrix} \tag{2}$$

that is generated by the MATLAB command **`gallery(5)`**. What is so exotic about this particular matrix? Well, enter it in your calculator or computer system of choice, and then use appropriate commands to show that:

- First, the characteristic equation of $\mathbf{A}$ reduces to $\lambda^5 = 0$, so $\mathbf{A}$ has the single eigenvalue $\lambda = 0$ of multiplicity 5.
- Second, there is only a single eigenvector associated with this eigenvalue, which thus has defect 4.

To seek a chain of generalized eigenvectors, show that $\mathbf{A}^4 \neq \mathbf{0}$ but $\mathbf{A}^5 = \mathbf{0}$ (the 5×5 zero matrix). Hence *any* nonzero 5-vector $\mathbf{u}_1$ satisfies the equation

$$(\mathbf{A} - \lambda\mathbf{I})^5\mathbf{u}_1 = A^5\mathbf{u}_1 = \mathbf{0}.$$

Calculate the vectors $\mathbf{u}_2 = \mathbf{A}\mathbf{u}_1$, $\mathbf{u}_3 = \mathbf{A}\mathbf{u}_2$, $\mathbf{u}_4 = \mathbf{A}\mathbf{u}_3$, and $\mathbf{u}_5 = \mathbf{A}\mathbf{u}_4$ in turn. You should find that $\mathbf{u}_5$ is nonzero, and is therefore (to within a constant multiple) the unique eigenvector $\mathbf{v}$ of the matrix $\mathbf{A}$. But can this eigenvector $\mathbf{v}$ you find possibly be independent of your original choice of the starting vector $\mathbf{u}_1 \neq \mathbf{0}$? Investigate this question by repeating the process with several different choices of $\mathbf{u}_1$.

Finally, having found a length 5 chain $\{\mathbf{u}_5, \mathbf{u}_4, \mathbf{u}_3, \mathbf{u}_2, \mathbf{u}_1\}$ of generalized eigenvectors based on the (ordinary) eigenvector $\mathbf{u}_5$ associated with the single eigenvalue $\lambda = 0$ of the matrix $\mathbf{A}$, write five linearly independent solutions of the 5-dimensional homogeneous linear system $\mathbf{x}' = \mathbf{A}\mathbf{x}$.

In the sections that follow we illustrate appropriate *Maple*, *Mathematica*, and MATLAB techniques to analyze the 4×4 matrix

$$\mathbf{A} = \begin{bmatrix} 35 & --12 & 4 & 30 \\ 22 & -8 & 3 & 19 \\ -10 & 3 & 0 & -9 \\ -27 & 9 & -3 & -23 \end{bmatrix} \tag{3}$$

of Problem 31 in Section 5.4 of the text. You can use any of the other problems there (especially Problems 23–30 and 32) to practice these techniques.

Using *Maple*

First we enter the matrix in (3):

```
with(linalg):
A := matrix(4,4, [ 35, -12,  4,  30,
                   22,  --8, 3,  19,
                  -10,   3,  0,  -9,
                  -27,   9, -3, -23 ] ):
```

Then we explore its characteristic polynomial, eigenvalues, and eigenvectors:

```
charpoly(A,lambda);
```

$$\lambda^4 - 4\lambda^3 + 6\lambda^2 - 4\lambda + 1$$

(that is, $(\lambda - 1)^4$)

```
eigenvals(A);
```

$$1, 1, 1, 1$$

```
eigenvects(A);
```

$$[1, 4, \{[0\ 1\ 3\ 0], [-1\ 0\ 1\ 1]\}]$$

Thus *Maple* finds only the two independent eigenvectors

```
w1 := matrix(4,1, [ 0,  1,  3,  0]):
w2 := matrix(4,1, [-1,  0,  1,  1]):
```

associated with the multiplicity 4 eigenvalue $\lambda = 1$, which therefore has defect 2. To explore the situation we set up the 4×4 identity matrix and the matrix $\mathbf{B} = \mathbf{A} - \lambda\mathbf{I}$:

```
Id := diag(1,1,1,1):
L = 1:
B := evalm( A - L*Id):
```

When we calculate $\mathbf{B}^2$ and $\mathbf{B}^3$,

```
B2 := evalm(B &* B);
B3 := evalm(B2 &* B);
```

we find that $\mathbf{B}^2 \neq 0$ but $\mathbf{B}^3 = 0$, so there should be a length 3 chain associated with $\lambda = 1$. Choosing

```
u1 := matrix(4,1,[1,0,0,0]);
```

we calculate the further generalized eigenvectors

```
u2 := evalm( B &* u1);
```

$$u2 := \begin{bmatrix} 34 \\ 22 \\ -10 \\ -27 \end{bmatrix}$$

and

```
u3 := evalm( B &* u2);
```

$$u3 := \begin{bmatrix} 42 \\ 7 \\ -21 \\ -42 \end{bmatrix}$$

Thus we have found the length 3 chain $\{\mathbf{u}_3,\ \mathbf{u}_2,\ \mathbf{u}_1\}$ based on the (ordinary) eigenvector $\mathbf{u}_3$. (To reconcile this result with *Maple*'s **eigenvects** calculation, you can check that $\mathbf{u}_3 + 42\mathbf{w}_2 = 7\mathbf{w}_1$.) Consequently four linearly independent solutions of the system $\mathbf{x}' = \mathbf{A}\,\mathbf{x}$ are given by

$$\begin{aligned}
\mathbf{x}_1(t) &= \mathbf{w}_1 e^t, \\
\mathbf{x}_2(t) &= \mathbf{u}_3 e^t, \\
\mathbf{x}_3(t) &= (\mathbf{u}_2 + \mathbf{u}_3 t)e^t, \\
\mathbf{x}_4(t) &= (\mathbf{u}_1 + \mathbf{u}_2 t + \tfrac{1}{2}\mathbf{u}_3 t^2)e^t.
\end{aligned}$$

Using *Mathematica*

First we enter the matrix in (3):

```
A = { { 35, -12,  4,  30 },
      { 22,  -8,  3,  19 },
      {-10,   3,  0,  -9 },
      {-27,   9, -3, -23 } };
```

Then we explore its characteristic polynomial, eigenvalues, and eigenvectors:

```
CharacteristicPolynomial[A, r]
```

```
               2       3     4
1 - 4 r + 6 r   - 4 r   + r
```

(that is, $(r-1)^4$)

```
Eigenvalues[A]
```

```
{1, 1, 1, 1}
```

```
Eigenvectors[A]
```

```
{{-3,-1,0,3}, {0,1,3,0}, {0,0,0,0}, {0,0,0,0}}
```

Thus *Mathematica* finds only the two independent (nonzero) eigenvectors

```
w1 = {-3,-1, 0, 3};
w2 = { 0, 1, 3, 0};
```

associated with the multiplicity 4 eigenvalue $\lambda = 1$, which therefore has defect 2. To explore the situation we set up the 4×4 identity matrix and the matrix $\mathbf{B} = \mathbf{A} - \lambda\mathbf{I}$:

```
Id = DiagonalMatrix[1,1,1,1];
L = 1;
B = A - L*Id;
```

When we calculate $\mathbf{B}^2$ and $\mathbf{B}^3$,

```
B2 = B.B
B3 = B2.B
```

we find that $\mathbf{B}^2 \neq 0$ but $\mathbf{B}^3 = 0$, so there should be a length 3 chain associated with $\lambda = 1$. Choosing

```
u1 = {{1},{0},{0},{0}}
```

we calculate

```
u2 = B.u1
```

{{34}, {22}, {-10}, {-27}}

```
u3 = B.u2
```

{{42}, {7}, {-21}, {-42}}

Thus we have found the length 3 chain $\{\mathbf{u}_3,\ \mathbf{u}_2,\ \mathbf{u}_1\}$ based on the (ordinary) eigenvector $\mathbf{u}_3$. (To reconcile this result with *Mathematica*'s **Eigenvectors** calculation, you can check that $\mathbf{u}_3 + 14\mathbf{w}_1 = -7\mathbf{w}_2$.) Consequently four linearly independent solutions of the system $\mathbf{x}' = \mathbf{A}\,\mathbf{x}$ are given by

$$\begin{aligned}
\mathbf{x}_1(t) &= \mathbf{w}_1 e^t, \\
\mathbf{x}_2(t) &= \mathbf{u}_3 e^t, \\
\mathbf{x}_3(t) &= (\mathbf{u}_2 + \mathbf{u}_3 t)e^t, \\
\mathbf{x}_4(t) &= (\mathbf{u}_1 + \mathbf{u}_2 t + \tfrac{1}{2}\mathbf{u}_3 t^2)e^t.
\end{aligned}$$

Using MATLAB

First we enter the matrix in (3):

```
A = [ 35   -12   4   30
      22    -8   3   19
     -10     3   0   -9
     -27     9  -3  -23 ];
```

Then we proceed to explore its characteristic polynomial, eigenvalues, and eigenvectors.

```
poly(A)
ans =
     1.0000    -4.0000     6.0000    -4.0000     1.0000
```

These are the coefficients of the characteristic polynomial, which hence is $(\lambda - 1)^4$. Then

```
[V, D] = eigensys(A)
V =
[ 1, 0]
[ 0, 1]
[-1, 3]
[-1, 0]
```

```
D =
[1]
[1]
[1]
[1]
```

Thus MATLAB finds only the two independent eigenvectors

```
w1 = [1  0  -1  -1]';
w2 = [0  1   3   0]';
```

associated with the single multiplicity 4 eigenvalue $\lambda = 1$, which therefore has defect 2. To explore the situation we set up the 4×4 identity matrix and the matrix $\mathbf{B} = \mathbf{A} - \lambda\mathbf{I}$:

```
Id = eye(4);
B = A - L*Id;
```

When we calculate $\mathbf{B}^2$ and $\mathbf{B}^3$,

```
B2 = B*B
B3 = B2*B
```

We find that $\mathbf{B}^2 \neq 0$ but $\mathbf{B}^3 = 0$, so there should be a length 3 chain associated with the eigenvalue $\lambda = 1$. Choosing the first generalized eigenvector

```
u1 = [1  0  0  0]';
```

we calculate the further generalized eigenvectors

```
u2 = B*u1
u2 =
    34
    22
   -10
   -27
```

and

```
u3 = B*u2
u3 =
    42
     7
   -21
   -42
```

Thus we have found the length 3 chain $\{\mathbf{u}_3,\ \mathbf{u}_2,\ \mathbf{u}_1\}$ based on the (ordinary) eigenvector $\mathbf{u}_3$. (To reconcile this result with MATLAB's **`eigensys`** calculation, you can check that $\mathbf{u}_3 - 42\mathbf{w}_1 = 7\mathbf{w}_2$.) Consequently four linearly independent solutions of the system $\mathbf{x}' = \mathbf{A}\mathbf{x}$ are given by

$$
\begin{aligned}
\mathbf{x}_1(t) &= \mathbf{w}_1 e^t, \\
\mathbf{x}_2(t) &= \mathbf{u}_3 e^t, \\
\mathbf{x}_3(t) &= (\mathbf{u}_2 + \mathbf{u}_3 t)e^t, \\
\mathbf{x}_4(t) &= (\mathbf{u}_1 + \mathbf{u}_2 t + \tfrac{1}{2}\mathbf{u}_3 t^2)e^t.
\end{aligned}
$$

Application 5.5
Automated Matrix Exponential Solutions

If $\mathbf{A}$ is an $n \times n$ matrix, then a computer algebra system can be used to calculate the fundamental matrix $e^{\mathbf{A}t}$ for the homogeneous linear system $\mathbf{x}' = \mathbf{A}\mathbf{x}$. Then Theorem 2 in Section 5.5 of the text says that the solution of the initial value problem

$$\mathbf{x}' = \mathbf{A}\mathbf{x}, \qquad \mathbf{x}(0) = \mathbf{x}_0 \tag{1}$$

is given by the (matrix) product

$$\mathbf{x}(t) = e^{\mathbf{A}t}\,\mathbf{x}_0. \tag{2}$$

In the paragraphs below we illustrate this approach by using *Maple*, *Mathematica*, and MATLAB to solve the initial value problem

$$\begin{aligned} x_1' &= 13x_1 + 4x_2, & x_1(0) &= 11 \\ x_2' &= 4x_1 + 7x_2, & x_2(0) &= 23. \end{aligned} \tag{3}$$

The success of this automated method and the usefulness of the results for a particular initial value problem may depend on whether sufficiently simple symbolic expressions for the eigenvalues and eigenvectors of the given matrix $\mathbf{A}$ can be found.

For a three-dimensional example to investigate, use this matrix exponential approach to solve the initial value problem

$$\begin{aligned} x_1' &= -149x_1 - 50x_2 - 154x_3, & x_1(0) &= 17 \\ x_2' &= 537x_1 + 180x_2 + 546x_3, & x_2(0) &= 43 \\ x_3' &= -27x_1 - 9x_2 - 25x_3, & x_3(0) &= 79. \end{aligned}$$

And here's a four-dimensional problem:

$$\begin{aligned} x_1' &= 4x_1 + x_2 + x_3 + 7x_4, & x_1(0) &= 15 \\ x_2' &= x_1 + 4x_2 + 10x_3 + x_4, & x_2(0) &= 35 \\ x_3' &= x_1 + 10x_2 + 4x_3 + x_4, & x_3(0) &= 55 \\ x_4' &= 7x_1 + x_2 + x_3 + 4x_4, & x_4(0) &= 75. \end{aligned}$$

If at this point you're having too much fun with matrix exponentials to stop, make up some problems of your own. For instance, choose any homogeneous linear system appearing in this chapter and experiment with differential initial conditions. The exotic 5×5 matrix **A** of the preceding 5.4 Application may suggest an interesting possibility.

Using *Maple*

First we define the coefficient matrix

```
with(linalg):
A := matrix(2,2, [13, 4, 4, 7] );
```

$$A := \begin{bmatrix} 13 & 4 \\ 4 & 7 \end{bmatrix}$$

and the initial vector

```
x0 := matrix(2,1, [11, 23]);
```

$$x0 := \begin{bmatrix} 11 \\ 23 \end{bmatrix}$$

for the initial value problem in (3). Then a fundamental matrix for the system $\mathbf{x}' = \mathbf{A}\mathbf{x}$ is given by

```
fundMatrix := exponential(A,t);
```

$$fundMatrix := \begin{bmatrix} \frac{1}{5}e^{(5t)} + \frac{4}{5}e^{(15t)} & -\frac{2}{5}e^{(5t)} + \frac{2}{5}e^{(15t)} \\ -\frac{2}{5}e^{(5t)} + \frac{2}{5}e^{(15t)} & \frac{4}{5}e^{(5t)} + \frac{1}{5}e^{(15t)} \end{bmatrix}$$

Hence Equation (2) gives the particular solution

```
solution := evalm(fundMatrix &* x0);
```

$$solution := \begin{bmatrix} -7e^{(5t)} + 18e^{(15t)} \\ 14e^{(5t)} + 9e^{(15t)} \end{bmatrix}$$

Thus the desired particular solution is given in scalar form by

$$x_1(t) := -7e^{5t} + 18e^{15t}, \qquad x_2(t) := 14e^{5t} + 9e^{15t}.$$

Using *Mathematica*

First we define the coefficient matrix

```
A = {{13, 4}, {4, 7}};
A // MatrixForm
```

```
13   4
4    7
```

and the initial vector

```
x0 = {{11}, {23}};
x0 // MatrixForm
```

```
11
23
```

for the initial value problem in (3). Then a fundamental matrix for the system $\mathbf{x}' = \mathbf{A}\mathbf{x}$ is given by

```
fundMatrix = MatrixExp[ A t ] // Expand
fundMatrix // MatrixForm
```

```
 5 t        15 t               5 t        15 t
E        4 E               -2 E        2 E
---- + -------             ------- + -------
 5        5                    5          5

    5 t        15 t            5 t     15 t
-2 E        2 E             4 E       E
------- + -------           ------ + -----
   5          5                5        5
```

Hence Equation (2) gives the particular solution

```
solution = fundMatrix . x0 // Expand
solution // MatrixForm
```

```
    5 t        15 t
-7 E     + 18 E

    5 t       15 t
14 E     + 9 E
```

Thus the desired particular solution is given in scalar form by

$$x_1(t) := -7e^{5t} + 18e^{15t},$$

$$x_2(t) := 14e^{5t} + 9e^{15t}.$$

Using MATLAB

First we define the coefficient matrix

```
A = [13 4; 4 7]

A =
    13     4
     4     7
```

and the initial vector

```
x0 = [11; 23]
x0 =
    11
    23
```

for the initial value problem in (3). Then a fundamental matrix for the system $\mathbf{x}' = \mathbf{A}\mathbf{x}$ is given by

```
syms t
fundMatrix = expm(A*t)

fundMatrix =

[ 4/5*exp(15*t)+1/5*exp(5*t),-2/5*exp(5*t)+2/5*exp(15*t)]
[-2/5*exp(5*t)+2/5*exp(15*t), 1/5*exp(15*t)+4/5*exp(5*t)]
```

Hence Equation (2) gives the particular solution

```
solution = fundMatrix*x0

solution =

[ 18*exp(15*t)-7*exp(5*t)]
[ 14*exp(5*t)+9*exp(15*t)]
```

Thus the desired particular solution is given in scalar form by

$$x_1(t) := -7e^{5t} + 18e^{15t},$$

$$x_2(t) := 14e^{5t} + 9e^{15t}.$$

Application 5.6
Automated Variation of Parameters

According to Eq. (28) in Section 5.6 of the text, the variation-of-parameters formula

$$\mathbf{x}(t) = e^{\mathbf{A}t}\left(\mathbf{x}_0 + \int_0^t e^{-\mathbf{A}s}\mathbf{f}(s)\,ds\right) \tag{1}$$

provides the solution to the nonhomogeneous initial value problem

$$\mathbf{x}' = \mathbf{A}\mathbf{x} + \mathbf{f}(t), \qquad \mathbf{x}(0) = \mathbf{x}_0 . \tag{2}$$

The formula in (1) constitutes a mechanical algorithm that encourages the use of a computer algebra system. In the sections that follow we illustrate this approach by applying *Maple*, *Mathematica*, and MATLAB to derive the solution

$$\mathbf{x}(t) = \frac{1}{14}\begin{bmatrix} (6+28t-7t^2)e^{-2t} + 92e^{5t} \\ (-4+14t+21t^2)e^{-2t} + 46e^{5t} \end{bmatrix} \tag{3}$$

of the initial value problem

$$\mathbf{x}' = \begin{bmatrix} 4 & 2 \\ 3 & -1 \end{bmatrix}\mathbf{x} - \begin{bmatrix} 15 \\ 4 \end{bmatrix} t e^{-2t}, \qquad \mathbf{x}(0) = \begin{bmatrix} 7 \\ 3 \end{bmatrix} \tag{4}$$

in Example 4 in the text. You can try this approach with Problems 17–34 in Section 5.6.

Using *Maple*

First we define the coefficient matrix

```
with(linalg):
A := matrix(2,2, [4,  2,
                  3, -1] ):
```

the initial vector

```
x0 := matrix(2,1, [7,
                   3]):
```

and the nonhomogeneous term

```
f := t -> matrix(2,1, [-15*t*exp(-2*t),
                        -4*t*exp(-2*t)]);
```

for the initial value problem in (4). The special matrix exponential function

```
expA := t-> exponential(A,t);
```

will simplify the notation a bit. We can now build up the solution in stages. First, the integrand matrix in the variation-of-parameters formula (1) is given by

```
integrand := evalm(expA(-s) &* f(s));
```

Next, *Maple*'s **map** function applies the integral function **int** element-wise to this matrix:

```
integral :=
simplify(map(int, integrand, s=0..t));
```

$$integral := \begin{bmatrix} -\frac{1}{2}t^2 + 2t\,e^{(-7t)} + \frac{2}{7}e^{(-7t)} - \frac{2}{7} \\ t\,e^{(-7t)} + \frac{1}{7}e^{(-7t)} + \frac{3}{2}t^2 - \frac{1}{7} \end{bmatrix}$$

Finally, the variation-of-parameters solution in (1) takes the form

```
solution :=
simplify(evalm(expA(t) &* (x0 + integral)));
```

$$solution := \begin{bmatrix} \frac{3}{7}e^{(-2t)} - \frac{1}{2}e^{(-2t)}t^2 + \frac{46}{7}e^{(5t)} + 2\,e^{(-2t)}t \\ \frac{23}{7}e^{(5t)} + e^{(-2t)}t - \frac{2}{7}e^{(-2t)} + \frac{3}{2}e^{(-2t)}t^2 \end{bmatrix}$$

which evidently is equivalent to the solution in Eq. (3) that was found manually in the text.

Using *Mathematica*

First we define the coefficient matrix

```
A = {{4,  2},
     {3, -1}};
```

the initial vector

```
x0 = {{7},
      {3}};
```

and the nonhomogeneous term

```
f[t_] := {{(-15*t)/E^(2*t)},
            {(-4*t)/E^(2*t)}}
```

for the initial value problem in (4). The special matrix exponential function

```
exp[A_] := MatrixExp[A]
```

will simplify the notation a bit. We can now build up the solution in stages. First, the integral matrix in the variation-of-parameters formula (1) is given by

```
integral =
Integrate[exp[-A*s] . f[s], {s, 0, t}] // Simplify
```

$$\begin{pmatrix} -\frac{t^2}{2} + 2e^{-7t}t + \frac{2}{7}\left(-1 + e^{-7t}\right) \\ \frac{3t^2}{2} + e^{-7t}t + \frac{1}{7}\left(-1 + e^{-7t}\right) \end{pmatrix}$$

Then the variation-of-parameters solution in (1) takes the form

```
solution :=
simplify(evalm(expA(t) &* (x0 + integral)));
```

$$\begin{pmatrix} \frac{1}{14}e^{-2t}\left(-7t^2 + 28t + 92e^{7t} + 6\right) \\ \frac{1}{14}e^{-2t}\left(21t^2 + 14t + 46e^{7t} - 4\right) \end{pmatrix}$$

which evidently is equivalent to the solution in Eq. (3) that was found manually in the text.

Using MATLAB

First we define the coefficient matrix

```
A = [4   2
     3  -1]
```

the initial vector

```
x0 = [7
      3]
```

and the nonhomogeneous function

```
syms s t
f = s*exp(-2*s)*[-15; -4]

f =
[ -15*s*exp(-2*s)]
[   -4*s*exp(-2*s)]
```

(as a function of s) for the initial value problem in (4). We can now build up the solution in stages. First, the integrand matrix in the variation-of-parameters formula (1) is given by

```
integrand = expm(-A*s)*f;
```

Next, the integral in (1) is given by

```
integral = int(integrand, s, 0,t);
```

Finally, the variation-of-parameters solution in (1) takes the form

```
solution = expm(A*t)*(x0 + integral);
solution = simple(solution)

solution =
[3/7/exp(t)^2-1/2/exp(t)^2*t^2+46/7*exp(t)^5+2/exp(t)^2*t]
[23/7*exp(t)^5+1/exp(t)^2*t-2/7/exp(t)^2+3/2/exp(t)^2*t^2]
```

When we pretty-print this solution,

```
pretty(solution)

[                            2                                   ]
[          1                t                   5          t     ]
[3/7  -------  -  1/2  -------  +  46/7 exp(t)   +  2  -------]
[           2                2                                  2]
[     exp(t)           exp(t)                          exp(t)   ]
[                                                                 ]
[                                                          2      ]
[                  5         t                1          t        ]
[ 23/7 exp(t)   +  -------  -  2/7  -------  +  3/2  -------   ]
[                        2                 2                2     ]
[                  exp(t)            exp(t)            exp(t)     ]
```

we see that it is equivalent to the solution in Eq. (3) that was found manually in the text.

Chapter 6

Nonlinear Systems and Phenomena

Application 6.1
Phase Plane Portraits and First-Order Equations

Consider a first-order differential equation of the form

$$\frac{dy}{dx} = \frac{G(x, y)}{F(x, y)}, \tag{1}$$

which may be difficult or impossible to solve explicitly. Its solution curves can nevertheless be plotted as trajectories of the corresponding autonomous two-dimensional system

$$\frac{dx}{dt} = F(x, y), \qquad \frac{dy}{dt} = G(x, y). \tag{2}$$

Most ODE plotters can routinely generate phase portraits for autonomous systems. Many of those appearing in Chapter 6 of the text were plotted using (as illustrated in the figure below) John Polking's **pplane** program (see J. Polking and D. Arnold, *Ordinary Differential Equations Using MATLAB* (2nd edition), Prentice Hall, 1999).

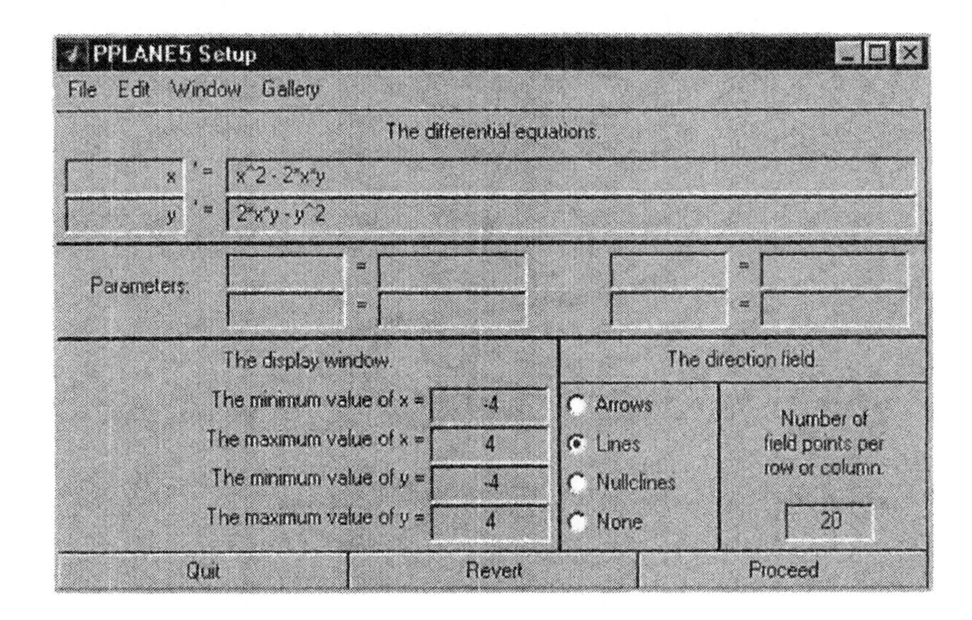

For example, to plot solution curves for the differential equation

$$\frac{dy}{dx} = \frac{2xy - y^2}{x^2 - 2xy} \tag{3}$$

we plot trajectories of the system

$$\frac{dx}{dt} = x^2 - 2xy, \qquad \frac{dy}{dt} = 2xy - y^2. \tag{4}$$

with the **pplane** setup of the preceding figure. The result is shown in Fig. 6.1.20 of the text. In the sections that follow the problems below we discuss the use of *Maple*, *Mathematica*, and MATLAB to construct such phase plane portraits.

Plot similarly some solution curves for the following differential equations.

1. $\dfrac{dy}{dx} = \dfrac{4x - 5y}{2x + 3y}$

2. $\dfrac{dy}{dx} = \dfrac{4x - 5y}{2x - 3y}$

3. $\dfrac{dy}{dx} = \dfrac{4x - 3y}{2x - 5y}$

4. $\dfrac{dy}{dx} = \dfrac{2xy}{x^2 - y^2}$

5. $\dfrac{dy}{dx} = \dfrac{x^2 + 2xy}{y^2 + 2xy}$

Now construct some examples of your own. Homogeneous functions like those in Problems 1 through 5 — rational functions with numerator and denominator of the same degree in x and y — work well. The differential equation

$$\frac{dy}{dx} = \frac{25x + y(1 - x^2 - y^2)(4 - x^2 - y^2)}{-25y + x(1 - x^2 - y^2)(4 - x^2 - y^2)} \tag{5}$$

of this form generalizes Example 6 in Section 6.1 of the text, but would be inconvenient to solve explicitly. Its phase portrait (Fig. 6.1.21) shows two periodic closed trajectories — the circles $r = 1$ and $r = 2$. Anyone want to try for three circles?

Using *Maple*

The **DEtools** package includes the **DEplot** function that can be used to construct a phase plane portrait for a 2-dimensional system of first-order differential equations. For instance, given the differential equations

```
deq1 := diff(x(t),t) = x^2 - 2*x*y:
deq2 := diff(y(t),t) = 2*x*y - y^2:
```

in (4), in order to plot the trajectories with initial conditions $x(0) = n$, $y(0) = n$ for $n = -4, -3, \ldots.., 2, 3, 4$, we need only enter the commands **with(DEtools):** and

```
DEplot([deq1,deq2], [x,y],
    t=-10..10, x=-5..5, y=-5..5,
   {[x(0)=-4,y(0)=-4],[x(0)=-3,y(0)=-3],
    [x(0)=-2,y(0)=-2],[x(0)=-1,y(0)=-1],
    [x(0)=1,y(0)=1],[x(0)=2,y(0)=2],
    [x(0)=3,y(0)=3],[x(0)=4,y(0)=4]},
    arrows = line, stepsize=0.02);
```

specifying the differential equations, the dependent variables, the t-range for each solution curve and the xy-plot ranges, a list of initial conditions for the desired solution curves, and the desired step size using the default Runge-Kutta method. The result is the collection of first- and third-quadrant trajectories shown in the figure below. You can investigate similarly the second- and fourth-quadrant trajectories of the system in (4).

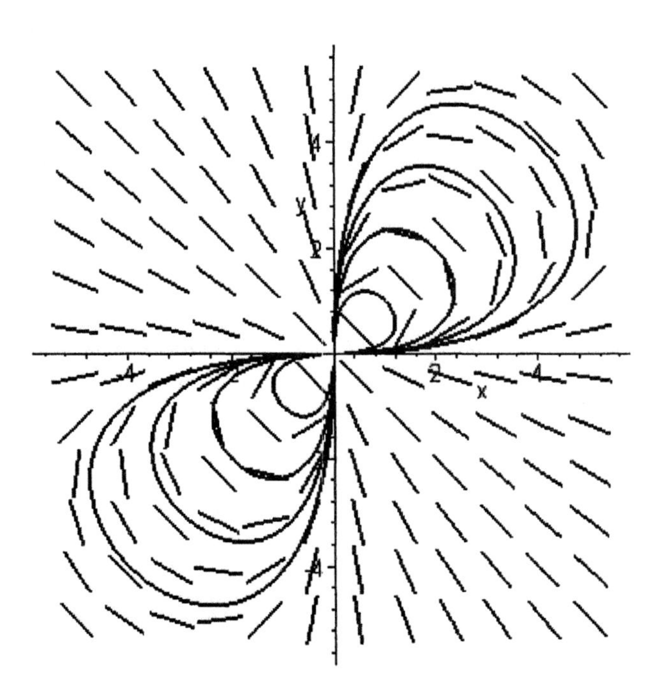

Using *Mathematica*

First we define the differential equations

```
deq1 = x'[t] == x[t]^2 - 2*x[t]*y[t];
deq2 = y'[t] == 2*x[t]*y[t] -  y[t]^2;
```

in (4). Then we can use the **NDSolve** function to solve the system numerically with given inital conditions.

```
soln =
NDSolve[ {deq1, deq2, x[0]==1, y[0]==1},
           {x[t],y[t]}, {t,-10,10} ]

{{x[t] -> InterpolatingFunction[{-10., 10.}, <>][t],
   y[t] -> InterpolatingFunction[{-10., 10.}, <>][t]}}
```

The corresponding solution curve is plotted with the commands

```
x = soln[[1,1,2]];
y = soln[[1,2,2]];
curve =
ParametricPlot[ {x,y}, {t,-10,10} ];
```

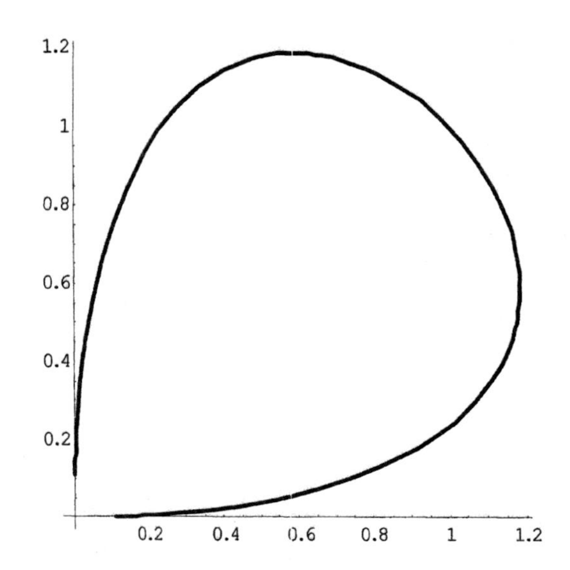

We can plot as many such solution curves as we like, and then display them simultaneously. For example, the following command creates a list (a **Table**) of solution curves corresponding to the initial conditions $x(0) = n,\ y(0) = n$ for $n = -4, -3, \ldots\ldots, 2, 3, 4$.

```
curve = Table[ n, {n,-4,4} ];
Do[Clear[x,y];
   soln =
   NDSolve[ {deq1,deq2,x[0] == n,y[0]== n},
            {x[t],y[t]}, {t,-10,10} ];
   x = soln[[1,1,2]];
   y = soln[[1,2,2]];
   curve[[n]] =
   ParametricPlot[ {x,y},
            {t,-10,10} ], {n,-4,4} ];
```

The command

```
Show[curve[[-4]],curve[[-3]],curve[[-2]],curve[[-1]],
     curve[[1]], curve[[2]], curve[[3]], curve[[4]],
     AspectRatio -> 1 ];
```

then displays these solution curves in a single phase plane portrait.

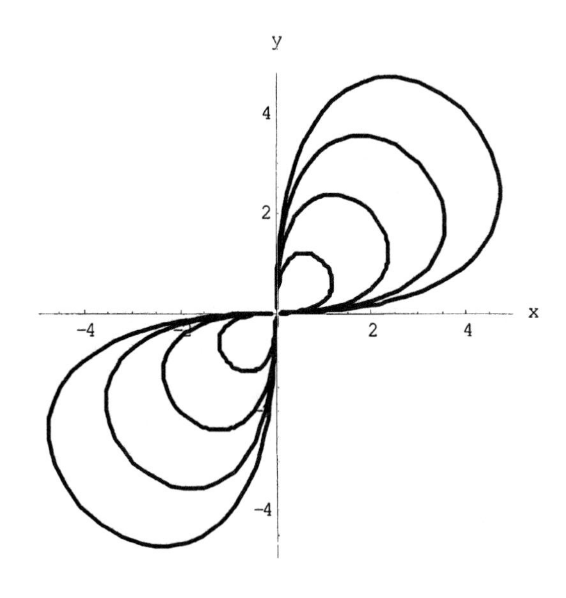

You can investigate similarly the second- and fourth-quadrant trajectories of the system in (4).

Using MATLAB

We will use the MATLAB ODE-solver **ode23** to construct a phase plane portrait for the system in (4). First we save the MATLAB m-file **yp.m** consisting of the lines

```
function     yp = yp(t,y)
%  yp.m
yp = y;
x = y(1);
y = y(2);
yp(1) = x.*x - 2*x.*y;
yp(2) = 2*x.*y - y.*y;
```

to define the system. Then the command

```
[t,y] = ode23('yp', [0,10], [4;4]);
```

generates a 2-column matrix **y** whose two columns are the x- and y-vectors for an approximate solution of the system on the interval $0 \le t \le 10$ with initial values $x(0) = y(0) = 4$. However, we may be surprised to find that the plot command

```
plot( y(:,1), y(:,2) )
```

then produces only half of the expected first-quadrant loop trajectory. (Try it and see for yourself!)

The other half of the desired "whole" trajectory is obtained by solving the solution system in (4) with the same initial condition, but on the interval $-10 \le t \le 0$. Actually, the whole loop corresponds to the infinite parameter interval $-\infty < t < \infty$ with the point $(x(t), y(t))$ on the trajectory approaching the origin as $t \to \pm\infty$, but the finite interval $-10 < t < 10$ suffices for graphical purposes.

We therefore proceed to generate directly a family of "whole" trajectories, solving the system numerically with the initial conditions $x(0) = y(0) = n$, twice for each of the successive values $n = -4, -3, \ldots\ldots, 2, 3, 4$. This is done by the simple **for**-loop

```
for n = -4 : 4
    [t,y] = ode23('yp', [0,10], [n;n] );
    plot(y(:,1),y(:,2),'b')
    hold on
    [t,y] = ode23('yp', [0,-10], [n;n] );
    plot(y(:,1),y(:,2),'b')
    end
```

which plots successively both "halves" of each first- and third-quadrant trajectory. The result is the figure shown on the next page. You can alter the initial conditions and t-intervals specified to investigate similarly the second- and fourth-quadrant trajectories of the system in (4).

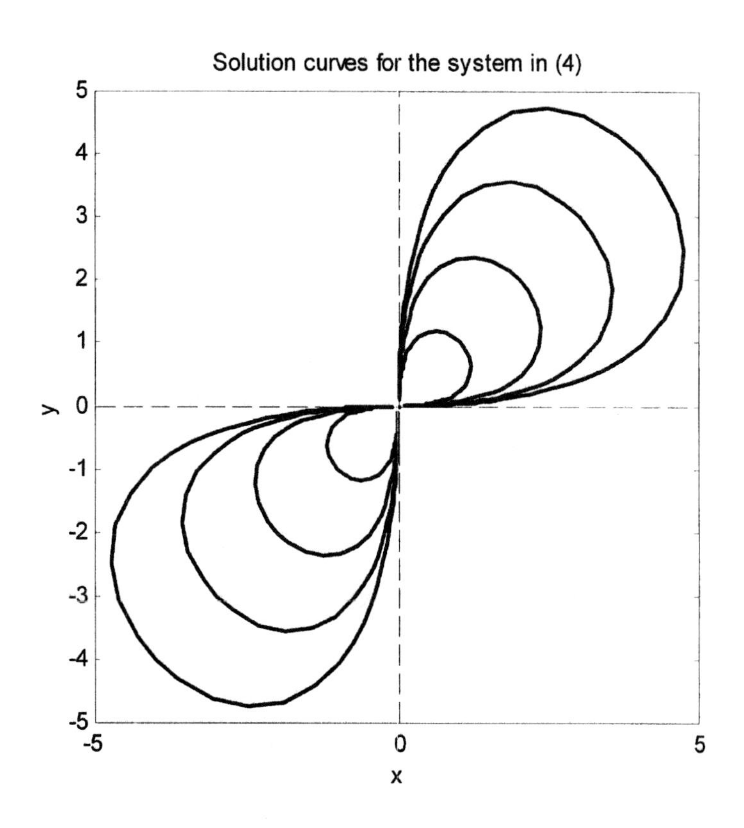
Solution curves for the system in (4)
y
x
5
4
3
2
1
0
-1
-2
-3
-4
-5
-5
0
5

Application 6.2
Phase Plane Portraits of Almost Linear Systems

Interesting and complicated phase portraits often result from simple nonlinear perturbations of linear systems. For instance, the figure below shows a phase plane portrait for the almost linear system

$$\begin{aligned}\frac{dx}{dt} &= -y\cos(x+y-1)\\ \frac{dy}{dt} &= x\cos(x-y+1).\end{aligned} \tag{1}$$

Among the seven critical points marked with dots, we see

- Apparent spiral points in the first and third quadrants of the *xy*-plane;
- Apparent saddle points in the second and fourth quadrants, plus another one on the positive *x*-axis;
- A critical point of undetermined character on the negative *y*-axis; and
- An apparently "very weak" spiral point at the origin -- meaning one that is approached very slowly as t increases or decreases (according as it is a sink or a source).

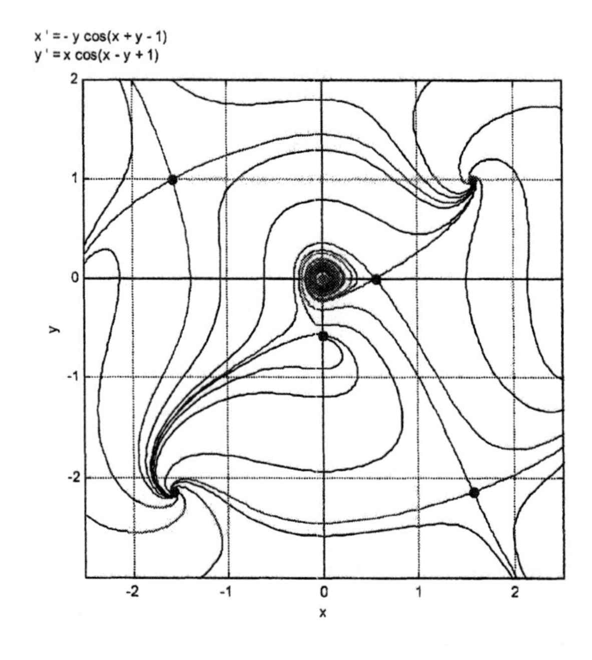

Some ODE software systems can automatically locate and classify critical points. For instance, Fig. 6.2.22 in the text shows a screen produced by John Polking's MATLAB **`pplane`** program (cited in the Section 6.1 application). It indicates that the fourth-quadrant critical point in the figure above has approximate coordinates (1.5708, –2.1416), and that the coefficient matrix of the associated linear system has the positive eigenvalue $\lambda_1 \approx 2.8949$ and the negative eigenvalue $\lambda_2 \approx -2.3241$. It therefore follows from Theorem 2 in Section 6.2 that this critical point is, indeed, a saddle point of the almost linear system
in (1).

With a general computer algebra system, you may have to do a bit of work yourself — or tell the computer precisely what to do — in order to find and classify a critical point. In the sections below, we illustrate this procedure using *Maple*, *Mathematica*, and MATLAB. Once the critical-point coordinates $a = 1.5708$, $b = -2.1416$ indicated above have been found, the substitution $x = u + a,\ \ y = v + b$ yields the translated system

$$\begin{aligned} \frac{du}{dt} &= (2.1416 - v)\cos(1.5708 - u - v) = f(u,v) \\ \frac{dv}{dt} &= (1.5708 + u)\cos(4.7124 + u - v) = g(u,v). \end{aligned} \tag{2}$$

If we substitute $u = v = 0$ in the Jacobian matrix

$$\mathbf{J} = \begin{bmatrix} \dfrac{\partial f}{\partial u} & \dfrac{\partial f}{\partial v} \\ \dfrac{\partial g}{\partial u} & \dfrac{\partial g}{\partial v} \end{bmatrix} \tag{3}$$

we get the coefficient matrix

$$\mathbf{A} = \begin{bmatrix} 2.1416 & 2.1416 \\ 1.5708 & -1.5708 \end{bmatrix} \tag{4}$$

of the linear system corresponding to the almost linear system in (2).

Alternatively, one can circumvent the translated system in (2) by looking at the Taylor expansions

$$\begin{aligned} f(x,y) &= D_x f(a,b)(x-a) + D_y f(a,b)(y-b) + \cdots \\ g(x,y) &= D_x g(a,b)(x-a) + D_y g(a,b)(y-b) + \cdots \end{aligned} \tag{5}$$

of the right-hand side functions in the original system (1), and retaining only the linear terms in this expansion. We see from (5) that

$$\mathbf{A} = \begin{bmatrix} D_x f(a,b) & D_y f(a,b) \\ D_x g(a,b) & D_y g(a,b) \end{bmatrix} \tag{6}$$

is the coefficient matrix of the linearization of the system (1) that results when we substitute $u = x - a,\ v = y - b$ and retain only the terms that are linear in u and v.

In any event, we can then use our computer algebra system to find the eigenvalues $\lambda_1 \approx 2.8949$ and $\lambda_2 \approx -2.3241$ of the matrix **A**, thereby verifying that the critical point $(1.5708, -2.1416)$ of (1) is, indeed, a saddle point.

Use a computer algebra system to find and classify similarly the other critical points of (1) indicated in the figure above. Then investigate similarly an almost linear system of your own construction. One convenient way to construct such a system is to start with a linear or almost linear system and insert sine or cosine factors resembling the ones in (1). For instance:

1. $x' = x\cos y$, $\quad y' = y\sin x$

2. $x' = -y + y^2\cos y$, $\quad y' = -x - x^2\sin x$

3. $x' = y\cos(2x + y)$, $\quad y' = -x\sin(x - 3y)$

4. $x' = -x - y^2\cos(x + y)$, $\quad y' = y + x^2\cos(x - y)$

Using *Maple*

After we enter the right-hand side functions in (1),

```
f := -y*cos(x+y-1):
g :=  x*cos(x-y-1):
```

we can proceed to solve numerically for a solution near $(1.5, -2)$:

```
soln :=
fsolve({f=0,g=0}, {x,y}, x=1..2, y=-3..-1);
```

{x = 1.570796327, y = -2.141592654}

Thus our critical point (a, b) is given approximately by

```
a := rhs(soln[1]);
b := rhs(soln[2]);
```

a := 1.570796327
b := -2.141592654

To classify this critical point, we proceed to calculate first the partial derivatives

```
fx := evalf(subs(x=a,y=b,diff(f,x))):
fy := evalf(subs(x=a,y=b,diff(f,y))):
gx := evalf(subs(x=a,y=b,diff(g,x))):
gy := evalf(subs(x=a,y=b,diff(g,y))):
```

evaluated at (a, b), and then the Jacobian matrix in (6):

```
with(linalg):
A := matrix(2,2, [fx,fy,gx,gy]);
```

Finally, its eigenvalues are given by

```
eigenvals(A);
                    2.894893108, -2.324096781
```

Thus the eigenvalues $\lambda_1 \approx 2.8949$ and $\lambda_2 \approx -2.3241$ are real with opposite signs, so the critical point (1.5708, –2.1416) is, indeed, a saddle point of the system in (1).

Using *Mathematica*

After we enter the right-hand side functions in (1),

```
f = -y*Cos[x+y-1];
g =  x*Cos[x-y+1];
```

we can proceed to solve numerically for a solution near $(1.5, -2)$:

```
soln =
FindRoot[{f == 0, g == 0}, {x,1.5}, {y,-2}]

{x -> 1.5708, y -> -2.14159}
```

Thus our critical point (a, b) is given approximately by

```
a = x /. soln
b = y /. soln

1.5708
-2.14159
```

To classify this critical point, we proceed to set up the Jacobian matrix in (6),

```
A = { {D[f,x], D[f,y]},
      {D[g,x], D[g,y]}} /. {x->a, y->b};
```

evaluated at (a, b), and calculate its eigenvalues

```
Eigenvalues[A]
{2.89489, -2.3241}
```

Thus the eigenvalues $\lambda_1 \approx 2.8949$ and $\lambda_2 \approx -2.3241$ are real with opposite signs, so the critical point (1.5708, –2.1416) is, indeed, a saddle point of the system in (1).

Using MATLAB

We want to solve the equations $f(x,y)=0,\ g(x,y)=0$ where f and g are the right-hand side functions in our system (1). However, the student edition of MATLAB does not include a function for the solution of systems of equations. Our strategy is therefore to use the MATLAB function **fminsearch** to minimize the function

$$h(x,y) = f(x,y)^2 + g(x,y)^2 = \left(-y\cos(x+y-1)\right)^2 + \left(x\cos(x-y+1)\right)^2$$

This function is defined as a function of the vector **v = [x;y]** by the m-file **h.m** consisting of the lines

```
function    z = h(v)
x = v(1);   y = v(2);
z = (-y*cos(x+y-1))^2 + (x*cos(x-y+1))^2;
```

Evidently a minimal point where $h(x,y)=0$ will be a critical point of the system in (1). Hence the commands

```
soln = fminsearch('h',[1.5;-2])
soln =
    1.5708
   -2.1416

a = soln(1);    b = soln(2);
```

yield the approximate critical point (1.5708, –2.1416). To classify this critical point, we proceed to set up the Jacobian matrix in (6). First we define the inline functions

```
f = '-y*cos(x+y-1)';
g = 'x*cos(x-y+1)';
```

calculate their partial derivatives

```
fx = diff(f,'x');      fy = diff(f,'y');
gx = diff(g,'x');      gy = diff(g,'y');
```

and evaluate these partial derivatives at the point (a, b):

```
syms x y
fx = subs(fx,  {x,y},  {a,b});
fy = subs(fy,  {x,y},  {a,b});
gx = subs(gx,  {x,y},  {a,b});
gy = subs(gy,  {x,y},  {a,b});
```

Then the eigenvalues of the Jacobian matrix

```
A = [fx  fy
     gx  gy];
```

are given by

```
eig(A)
ans =
     2.8948
    -2.3241
```

Thus the eigenvalues $\lambda_1 \approx 2.8948$ and $\lambda_2 \approx -2.3241$ are real with opposite signs, so the critical point (1.5708, –2.1416) is, indeed, a saddle point of the system in (1).

Application 6.3
Predator-Prey and Your Own Game Preserve

The closed trajectories in the figure below represent periodic solutions of a typical predator-prey system, but provide no information as to the actual periods of the population oscillations these solutions describe.

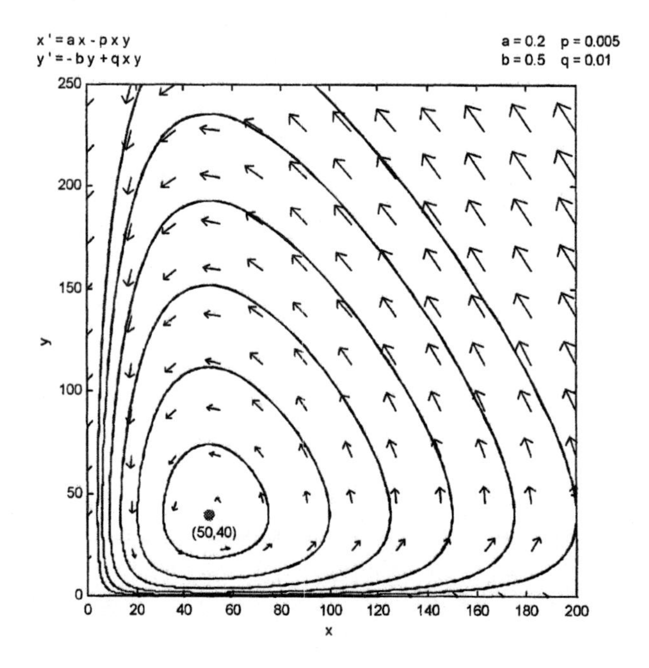

The period P of a particular solution $(x(t), y(t))$ can be gleaned from the graphs of x and y as functions of t. The figure on the next page shows these graphs for the particular solution satisfying the initial conditions $x(0) = 70,\ y(0) = 30$. The labeled period P indicates how the period with which the x- and y-populations oscillate can be measured — at least approximately — on such a figure.

Investigation 1

You own a large forest hunting preserve that you originally stocked with F_0 foxes and R_0 rabbits on January 1, 2007. The following differential equations model the numbers $R(t)$ of rabbits and $F(t)$ of foxes t months later.

$$\begin{aligned} \frac{dR}{dt} &= 0.01\,p\,R - 0.0001\,a\,RF \\ \frac{dF}{dt} &= -0.01\,q\,F - 0.0001\,b\,RF \end{aligned} \qquad (1)$$

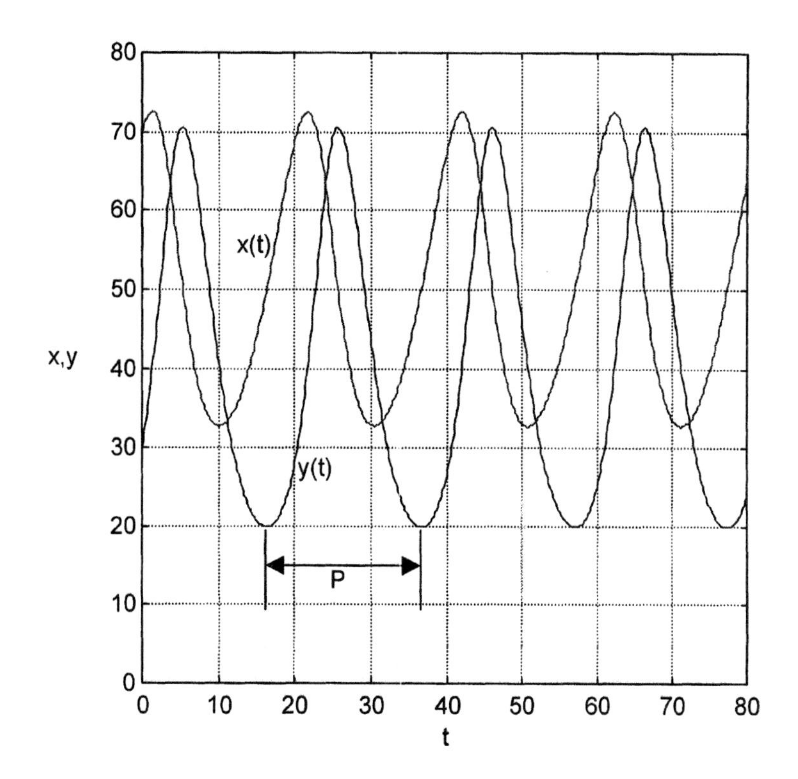

where p and q are the two largest digits (with $p < q$), and a and b are the two smallest nonzero digits (with $a < b$) in your student ID number.

The numbers of foxes and rabbits will oscillate periodically, out of phase with each other (like the functions $x(t)$ and $y(t)$ in the figure above). Pick your initial numbers F_0 of foxes and R_0 of rabbits -- perhaps several hundred of each -- so that the resulting solution curve in the *RF*-plane is a fairly eccentric closed curve. (The eccentricity may be increased if you start with a largish number of rabbits and a smallish number of foxes, as any hunting preserve owner would naturally do -- since foxes eat rabbits.)

Your task then is to determine

- The period of oscillation of the rabbit and fox populations;
- The maximum and minimum numbers of rabbits, and the calendar dates on which they first occur; and
- The maximum and minimum numbers of foxes, and the calendar dates on which they first occur.

With computer software that can plot both *RF*-trajectories and *tR*- and *tF*-solution curves like those above, you can "zoom in" graphically on the points whose coordinates provide the requested information.

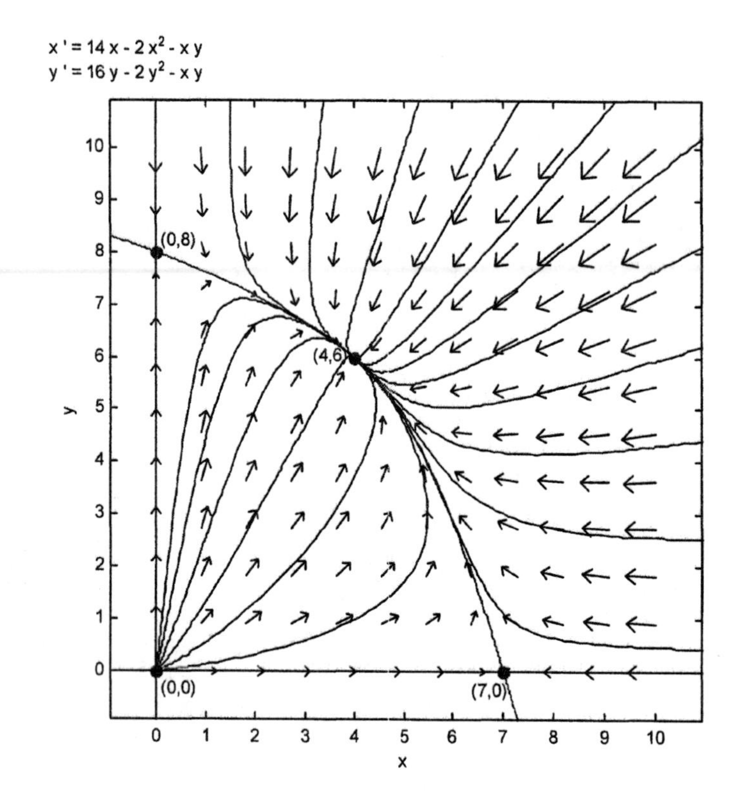

Investigation 2

For a more general ecological system to investigate, let a, b, c, d be the four smallest nonzero digits (in any order) and m, n the two largest digits in your student ID number. Then consider the system

$$\frac{dx}{dt} = x(P - ax \pm by), \qquad \frac{dy}{dt} = y(Q \pm cx - dy) \qquad (2)$$

Where $P = ma - (\pm nb)$ and $Q = nd - (\pm mc)$, with the same choice of plus/minus signs in dx/dt and P and (independently) in dy/dt and Q — so that (m, n) is a critical point of the system. Then use the methods of the Section 6.1 application to plot a phase plane portrait for this system in the first quadrant of the xy-plane. In particular, determine the long-term behavior (as $t \to \infty$) of the two populations, in terms of their initial populations $x(0) = x_0$ and $y(0) = y_0$. For instance, the figure above shows a phase plane portrait for the system

$$\frac{dx}{dt} = x(14 - 2x - y), \qquad \frac{dy}{dt} = y(16 - x - 2y).$$

We see a nodal source at $(0, 0)$, a nodal sink at $(4, 6)$, and saddle points at $(7, 0)$ and $(0, 8)$. It follows that, if x_0 and y_0 are both positive, then $x(t) \to 4$ and $y(t) \to 6$ as $t \to \infty$.

In the sections that follow we use the simple predator-prey system

$$\frac{dx}{dt} = x - xy, \qquad \frac{dy}{dt} = -y + xy. \tag{3}$$

to illustrate the *Maple*, *Mathematica*, and MATLAB techniques needed for these investigations.

Using *Maple*

To plot a solution curve for the system in (3) we need only load the **DEtools** package and use the **DEplot** function. For instance, if we first define the differential equations

```
deq1 :=  diff(x(t),t) =  x - x*y:
deq2 :=  diff(y(t),t) = -y + x*y:
```

then the commands

```
with(DEtools):
DEplot([deq1,deq2], [x,y], t=0..25,
       {[x(0)=1, y(0)=3]}, stepsize=0.1,
       linecolor=blue, arrows=none);
```

plot the xy-solution curve with initial conditions $x(0) = 1$, $y(0) = 3$ on the interval $0 \le t \le 25$ with step size $h = 0.1$. Next, the command

```
DEplot([deq1,deq2], [x,y], t=0..25,
       {[x(0)=1, y(0)=3]}, stepsize=0.1,
       scene = [t,x], linecolor=blue, arrows=none);
```

plots the corresponding tx-solution curve, on which the approximate period of oscillation of the prey population can be measured.

Using *Mathematica*

To plot a solution curve for the system in (3) we need only define the differential equations

```
deq1 = x'[t] ==  x[t] - x[t]*y[t];
deq2 = y'[t] == -y[t] + x[t]*y[t];
```

and then use **NDSolve** to integrate numerically. For instance, the command

```
soln = NDSolve[ {deq1,deq2, x[0]==1, y[0]==3},
                {x[t],y[t]}, {t,0,25} ]
```

yields an approximate solution on the interval $0 \le t \le 25$ satisfying the initial conditions $x(0) = 1,\ y(0) = 3$. Then the command

```
ParametricPlot[
    Evaluate[{x[t],y[t]} /. soln], {t,0,25}]
```

plots the corresponding *xy*-solution curve, and the command

```
Plot[Evaluate[ x[t] /. soln ], {t,0,25}]
```

plots the corresponding *tx*-solution curve, on which the approximate period of oscillation of the prey population can be measured.

Using MATLAB

To plot a solution curve for the system in (3) we need only define the system by means of the m-file

```
function  yp = predprey(t,y)
% predprey.m
yp = y;
x = y(1);
y = y(2);
yp(1)  = x - x.*y;
yp(2)  = -y + x.*y;
```

and then use **ode23** to integrate numerically. For instance, the command

```
[t,x] = ode23('predprey', [0:0.1:25], [1;3]);
```

yields an approximate solution on the interval $0 \le t \le 25$ satisfying the initial conditions $x(0) = 1,\ y(0) = 3$. Then the command

```
plot(x(:,1), x(:,2))
```

plots the corresponding *xy*-solution curve, and the command

```
plot(t, x(:,1))
```

plots the corresponding *tx*-solution curve, on which the approximate period of oscillation of the prey population can be measured.

Application 6.4
The Rayleigh and van der Pol Equations

The British mathematical physicist Lord Rayleigh (John William Strutt, 1842–1919) introduced an equation of the form

$$mx'' + kx = ax' - b(x')^3 \tag{1}$$

(with non-linear velocity damping) to model the oscillations of a clarinet reed. With $y = x'$ we get the autonomous system

$$x' = y, \qquad y' = \frac{1}{m}\left(-kx + ay - by^3\right). \tag{2}$$

A typical phase portrait for this system is shown in Fig. 6.5.15 in the text, where we see outward and inward spiral trajectories converging to a "limit cycle" solution that corresponds to periodic oscillations of the reed.

The figure below shows an *xy*-trajectory with parameter values $m = 2$ and $k = a = b = 1$. The period $P \approx 9.2$ (and hence the frequency) of these oscillations can be measured (approximately) as indicated on the *tx*-solution curve plotted at the top of the next page. This period of oscillation depends only on the parameters $m,\ k,\ a,$ and b in Eq. (1), and is independent of the initial conditions (why?).

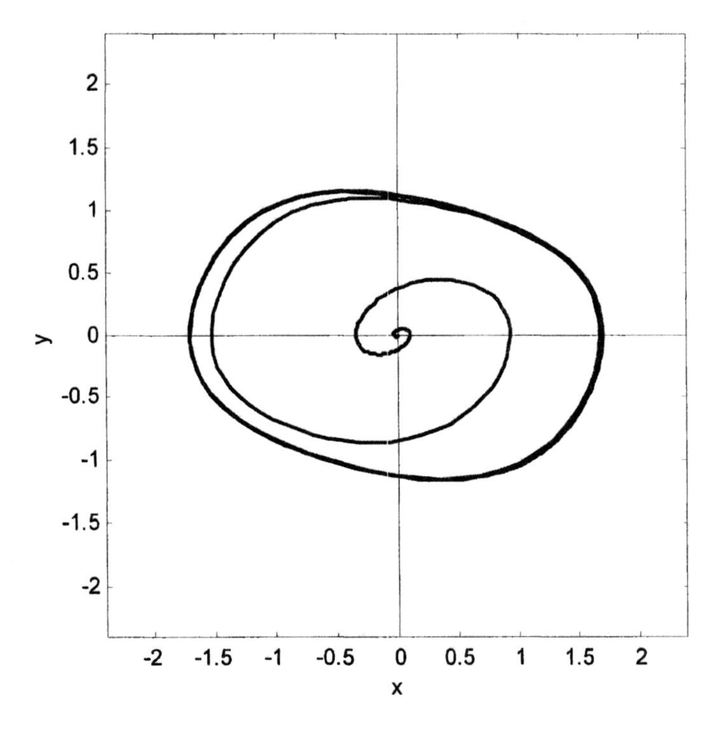

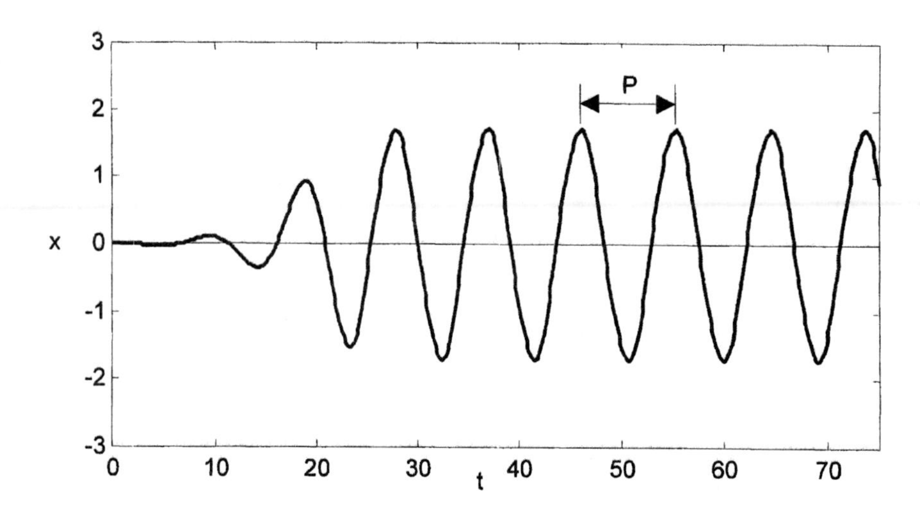

Choose your own parameters m, k, a, and b (perhaps the least four nonzero digits in your student ID number), and use an available ODE plotting utility to plot both xy-trajectories and tx-solution curves for the resulting Rayleigh equation. Change *one* of your parameters (perhaps m) to see how the amplitude and frequency of the resulting periodic oscillations are altered.

Van der Pol's Equation

Figure 6.4.17 in the text shows a simple RLC circuit in which the usual (passive) resistance R has been replaced by an active element (such as a vacuum tube or semiconductor) across which the voltage drop V is given by a known function $f(I)$ of the current I. Of course, $V = f(I) = RI$ for a standard resistor. If we substitute $f(I)$ for RI in the familiar RLC-circuit equation $LI'' + RI' + I/C = 0$ of Section 3.7, we get the second-order equation

$$LI'' + f'(I)\,I' + I/C = 0. \tag{3}$$

In a 1924 study of oscillator circuits in early commercial radios, Balthasar van der Pol (1889-1959) assumed the voltage drop to be given by a nonlinear function of the form $f(I) = bI^3 - aI$, which with Eq. (3) becomes

$$LI'' + (3bI^2 - a)I' + I/C \;=\; 0. \tag{4}$$

This equation is closely related to Rayleigh's equation, and has phase portraits resembling the one shown for Rayleigh's equation. Indeed, differentiation of the second equation in (2) and the substitution $x' \;=\; y$ yield the equation

$$my'' + (3by^2 - ay)y' + ky \;=\; 0 \tag{5}$$

which has the same form as Eq. (4).

If we denote by τ the time variable in Eq. (4) and make the substitutions $I = px,\ t = \tau/\sqrt{LC}$, the result is

$$\frac{d^2x}{dt^2} + \left(3bp^2x^2 - a\right)\sqrt{\frac{C}{L}}\frac{dx}{dt} + x = 0.$$

With $p = \sqrt{a/(3b)}$ and $\mu = a\sqrt{C/L}$ this gives the standard form

$$x'' + \mu(x^2 - 1)x' + x = 0 \tag{6}$$

of *van der Pol's equation.*

For every nonnegative value of the parameter μ, the solution of van der Pol's equation with $x(0) = 2,\ x'(0) = 0$ is periodic, and the corresponding phase plane trajectory is a limit cycle to which the other trajectories converge (as illustrated in Fig. 6.4.15 in the text). It will be instructive for you to solve van der Pol's equation numerically and plot this periodic trajectory for a selection of values from $\mu = 0$ to $\mu = 1000$ or more. With $\mu = 0$ it is a circle of radius 2 (why?). Figure 6.4.18 in the text shows the periodic trajectory with $\mu = 1$, and Fig. 6.4.19 shows the corresponding $x(t)$ and $y(t)$ solution curves.

You might also plot other trajectories that are "attracted" from within and from without by the limit cycle (like those shown in Fig. 6.4.18). The origin looks like a spiral point in Fig. 6.4.18. Indeed, show that (0,0) is a spiral source for van der Pol's equation if $0 < \mu < 2$, but is a nodal source if $\mu \geq 2$.

In the sections that follow we illustrate the *Maple*, *Mathematica*, and MATLAB techniques needed for these investigations.

Using *Maple*

To plot a solution curve of the Rayleigh system in (2) we need only load the **DEtools** package and use the **DEplot** function. For instance, if we first specify the parameter values

```
m := 2:      k := 1:
a := 1:      b := 1:
```

and define the differential equations

```
deq1 :=     diff(x(t),t) = y:
deq2 :=   m*diff(y(t),t) = -k*x + a*y - b*y^3:
```

then the commands

```
with(DEtools):
DEplot([deq1,deq2], [x,y],
       t=0..75, x=-3..3, y=-3..3,
       {[x(0)=0.01, y(0)=0]}, stepsize=0.1,
       linecolor=blue, arrows=none);
```

plot the outward spiraling xy-trajectory with initial conditions $x(0)=0.01,\quad y(0)=0,$ on the interval $0 \le t \le 75$ with step size $h=0.1$. Next, the command

```
DEplot([deq1,deq2], [x,y], t=0..75,
       {[x(0)=0.01, y(0)=0]}, stepsize=0.1,
       scene = [t,x], linecolor=blue, arrows=none);
```

plots the corresponding tx-solution curve, on which the approximate period of oscillation can be measured.

Using *Mathematica*

To plot a solution curve for the Rayleigh system in (2) we need only specify the parameter values

```
m = 2;     k = 1;
a = 1;     b = 1;
```

define the differential equations

```
deq1 =     x'[t] ==  y[t];
deq2 =   m*y'[t] == -x[t] + a*y[t] - b*y[t]^3;
```

and then use **NDSolve** to integrate numerically. For instance, the command

```
soln = NDSolve[ {deq1,deq2, x[0]==0.01, y[0]==0},
                {x[t],y[t]}, {t,0,75} ]
```

yields an approximate solution on the interval $0 \le t \le 75$ satisfying the initial conditions $x(0)=0.01,\ y(0)=0$. Then the command

```
ParametricPlot[
     Evaluate[{x[t],y[t]} /. soln], {t,0,75}]
```

plots the corresponding xy-trajectory, and the command

```
Plot[Evaluate[ x[t] /. soln ], {t,0,75}]
```

plots the corresponding tx-solution curve, on which the approximate period of oscillation of the prey population can be measured.

Using MATLAB

To plot solution curves for the van der Pol system

$$x' = y, \qquad y' = -\mu(x^2 - 1)y - x \tag{7}$$

corresponding to Eq. (6), we first define the system by means of the m-file

```
function  yp = vanderpol(t,y)
% vanderpol.m
yp = y;
x = y(1);
y = y(2);
mu = 2;
yp(1)  = y;
yp(2)  = -mu*(x^2 - 1)*y - x;
```

Note that the value $\mu = 2$ is specified in this function file. Then we use **ode45** to integrate numerically. For instance, the command

```
[t,y] = ode45('vanderpol', [0:0.04:20], [2; 0]);
```

yields an approximate solution on the interval $0 \le t \le 20$ satisfying the initial conditions $x(0) = 2,\ y(0) = 0$. Then the command

```
plot(y(:,1), y(:,2))
```

plots the corresponding xy-solution curve shown at the top of the next page. Finally, the command

```
plot(t,y(:,1), t,y(:,2) )
```

plots the corresponding tx- and ty-solution curves, on which the approximate period of oscillation can be measured. This plot is shown at the bottom of the next page.

When μ is large, van der Pol's equation is quite "stiff" and the periodic trajectory is more eccentric as illustrated in Fig. 6.4.20 in the text, which was plotted using MATLAB's stiff ODE solver **ode15s**. Indeed, the ability to plot Fig. 6.4.20 accurately — and especially the corresponding $x(t)$ and $y(t)$ solution curves analogous to those in Fig. 6.4.19 — is a good test of the robustness of your computer system's ODE solver.

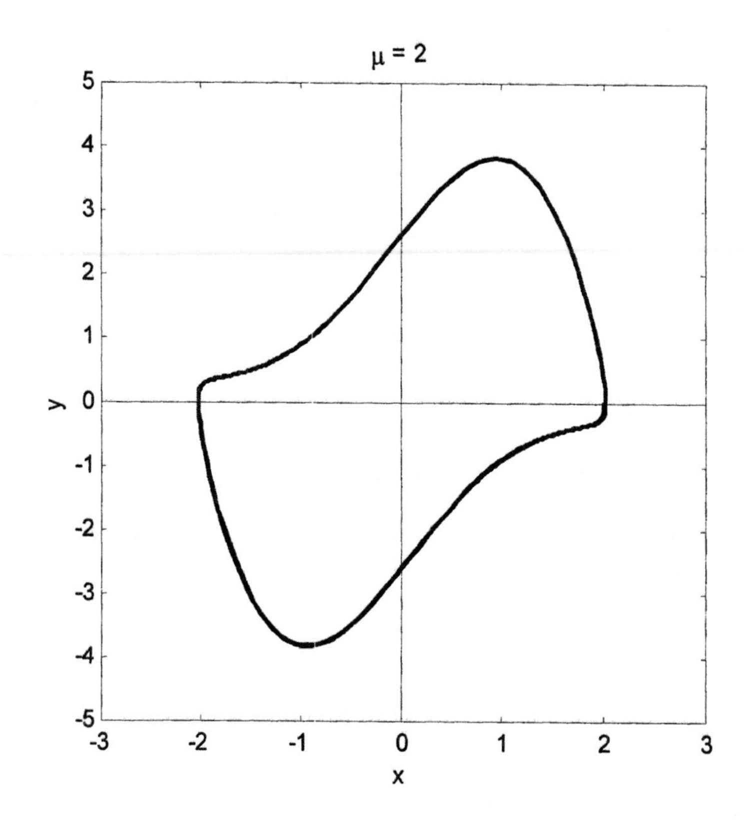
μ = 2
5
4
3
2
1
0
-1
-2
-3
-4
-5
y
-3
-2
-1
0
1
2
3
x

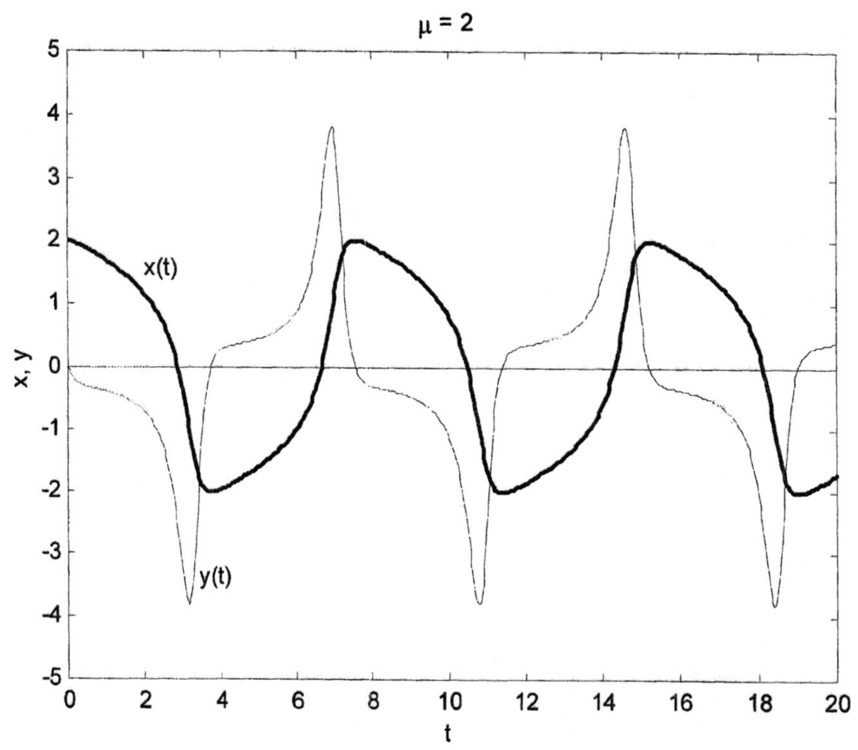
μ = 2
x(t)
y(t)
x, y
0
2
4
6
8
10
12
14
16
18
20
t

Application 6.5A
Period Doubling and Pitchforks

We list first some programs that can be used to investigate periodic cycles for the iteration function $g(x) = rx(1-x)$, as illustrated in Figures 6.5.2–6.5.6 in the text. The following BASIC program asks the user to specify the value of the parameter r and the length k of the blocks of iterates to display. After a thousand initial iterations to stabilize, 5 blocks of k iterates each are displayed for inspection.

```
100 'BASIC Program PERIODS
110 '
120 'For the iteration  x = rx(1 - x)
130 '
140  DEFDBL R,X
150  INPUT "Value of r"; R
160  INPUT "Print in blocks of k  =  "; K
170  P$  =  "#.####     "
180  X  =  .5                        'Initial seed
190 '
200  FOR I  =  1 TO 1000              '1000 initial
210     X  =  R*X*(1 - X)            'iterations to
220  NEXT I                          'stabilize.
230 '
240  FOR  J  =  1  TO  5
250     FOR  I  =  1  TO  K
260        X  =  R*X*(1 - X)         'Final iterations
270        PRINT USING P$; X;
280     NEXT I
290     IF K <> 8 THEN PRINT
300  NEXT J
310  END
```

The following table gives some interesting values of r (and the resulting periods) to try.

r	Period
2.75	1
3.25	2
3.50	4
3.55	8
3.565	16
3.57	Chaos
3.84	3
3.845	6
3.848	12

Using *Maple*

To check out the period 4 iterates with $r = 3.50$, we first carry out 1000 iterations to stabilize.

```
r := 3.5:
x := 0.5:
for i from 1 to 1000 do
        x := r*x*(1-x);
        od:
```

Then we can display two blocks of $k = 4$ iterates each with the commands

```
k := 4:
for i from 1 to 2*k do
         x := r*x*(1-x);
         od;
```

x := .8749972638
x := .3828196825
x := .8269407063
x := .5008842110

x := .8749972638
x := .3828196825
x := .8269407063
x := .5008842110

Thus we see the alleged cycles of length $k = 4$.

Using *Mathematica*

We first define the iteration function

```
g[x_] := r x (1 - x)
```

Then the concise code

```
r = 3.5;
x0 = 0.5;
NestList[g, x0, 1012];
```

calculates the first 1012 iterates of g with parameter value $r = 3.5$ starting with $x_0 = 0.5$, and the commands

```
k = 4;
N[Partition[Drop[%, 1000],k] // TableForm, k]

0.5009   0.875   0.3828   0.8269
0.5009   0.875   0.3828   0.8269
0.5009   0.875   0.3828   0.8269
```

display the last 12 of these iterates in blocks of 4, thereby exhibiting the alleged cycles of length 4.

Using MATLAB

To check out the period 4 iterates with $r = 3.50$, we first carry out 500 iterations to stabilize.

```
r = 3.5;
x = 0.5;
for i = 1 : 500
     x = r*x*(1-x);
     end
```

Then we can display 3 blocks of $k = 4$ iterates each with the commands

```
k = 4;
X = [];
for i = 1 : 3*k
      x = r*x*(1-x);
      X = [X, x];
      end
reshape(X, k,3)'
ans =
    0.8750    0.3828    0.8269    0.5009
    0.8750    0.3828    0.8269    0.5009
    0.8750    0.3828    0.8269    0.5009
```

Thus we see the alleged cycles of length 4.

For your further investigations, use pitchfork diagrams to look for other interesting cycles, and verify their apparent periods by appropriate iterative computations. For instance, you should find a cycle with period 10 between $r = 3.60$ and $r = 3.61$, and one with period 14 between $r = 3.59$ and $r = 3.60$. Can you find cycles with period 5 and 7? If so, look for subsequent period-doubling. A single run of of a typical pitchfork program requires several hundred thousand iterations, so it will help if you have a fast computer (or one you can leave running overnight).

Pitchfork Diagram Programs

The following BASIC program can be used to plot pitchfork diagrams as shown in Figures 6.5.8–6.5.9 in the text. As written, it runs well in Borland TurboBasic, but may have to be fine-tuned to run in other dialects of BASIC.

```
100 'Program PICHFORK
110 '
120 'Exhibits the period-doubling toward chaos
130 'generated by the Verhulst iteration
140 '
150 '        x  =  rx(1 - x)
160 '
170 'as the growth parameter  r  is increased
180 'in the range from about  3  to about  4.
190 '
200  DEFDBL H,K,R,X
210  DEFINT I,J,M,N,P,Q
220  INPUT "Rmin,Rmax"; RMIN, RMAX    'Try 2.8 and 4.0
230  INPUT "Xmin,Xmax"; XMIN, XMAX    'Try  0  and  1
240 '
250  KEY OFF  : CLS
260 'SCREEN 1  :  N  =  319     'For med resolution
270  SCREEN 2  :  N  =  639     'For hi  resolution
280  M  =  200                  'Hor rows for either
290  H  =  (RMAX - RMIN)/N
300  K  =  (XMAX - XMIN)/M
310 '
320  LINE (0,0) - (N,0)         'Draws a box
330  LINE - (N,199)             'around
340  LINE - (0,199)             'the screen
350  LINE - (  0,0)
360 '
370  FOR P  =  1 TO 9                'Tick marks on
380      Q  =  (P*(N+1)/10)  - 1     'top and bottom
390      LINE (Q,0) - (Q,5)          'of box
400      LINE (Q,195) - (Q,199)
410  NEXT P
420 '
430  FOR J  =  0 TO  N          'Jth vertical column
440      R  =  RMIN + J*H       'of pixels on screen
450      X  =  .5
460      FOR P  =  0 TO 1000    'These iterations
470          X  =  R*X*(1-X)    'to settle down.
480      NEXT P
```

```
490       FOR Q  =  0 TO 250     'These iterations
500           X  =  R*X*(1-X)    'are recorded.
510           I  =  INT((X - XMIN)/K)
520           I  =  200 - I
530           IF (0< = I) AND (I<200) THEN PSET (J,I)
540       NEXT Q
550   NEXT J
560 '
570   WHILE INKEY$  =  ""                'Press a key when
580   WEND                               'finished looking.
590   SCREEN 0  :  CLS  :  KEY ON
600   END
```

Figures 6.5.6 and 6.5.7 in the first edition of the text — corresponding to Figs. 6.5.8 and 6.5.9 in the third edition — were plotted using the following *Mathematica* program, a slight elaboration of one found on page 102 of T. Gray and J. Glynn, **Exploring Mathematics with *Mathematica***, Addison-Wesley, 1991. However, this program runs so slowly that it was necessary to leave a now-obsolete 68040 Macintosh running overnight some years ago.

```
g[x_] : =  r x (1 - x);
Clear[r];
a  =  2.8;  b  =  4.0;    (* r-range for Fig 6.5.6 *)
c  =  0;    d  =  1;      (* x-range               *)
m  =  250;                (* no of x-points        *)
n  =  500;                (* no of r-values        *)
ListPlot[
     Flatten[Table[
          Transpose[{
               Table[r, {m+1}],
                    NestList[g, Nest[q, 0.5, 2m], m]
                    }],
               {r, a, b, (b-a)/n}
                         ],
                              1],
     PlotStyle -> PointSize[0.001],
     PlotRange -> {{a,b},{c,d}},
     AspectRatio -> 0.75,
     Frame -> True,
     AxesLabel -> {"r","x"} ]
```

The MATLAB program on the next page runs much faster on a comparable computer, and is a good deal easier to understand. It required only about 35 seconds to generate Fig. 6.5.9 using a 266 MHz Pentium II computer (which now itself is obsolete).

```
% pichfork.m
% pitchfork diagram script for Figures 6.5.8 and 6.5.9

hold off
m  =  400;              % no of r-subintervals
n  =  400;              % no of x-subintervals

a  =  3.8;   b  =  3.9;        % r-range for 6.5.9
dr  =  (b - a)/m;
R  =  a+dr/2 : dr : b;         % vector of r-values
c  =  0;      d  =  1;         % x-range
dx  =  (d - c)/n;
X  =  c+dx/2 : dx : d;         % vector of x-values
[rr,xx]  =  meshgrid(R,X);     % matrices of r- and x-coords
                               % of grid points in rx-rect
C = zeros(m,n);
for j  =  1 : m               % Cycle through r-values
    r  =  a - dr/2 + j*dr;
    x  =  0.5;                % Initialize x-value
    for k  =  1:1000          % 1000 iterations to stabilize
        x  =  r*x*(1-x);
        end
    for k  =  1:1000
        x  =  r*x*(1-x);
         i = ceil(x/dx);
         C(i,j) = 1;         % lattice point to plot
         end
    end

C = C + 1;
C = flipud(C);
image(R,X,C)
colormap([1 1 1; 0 0 0])
axis square
```

Application 6.5B
Period Doubling and Chaos in Mechanical Systems

The first objective of this section is the application of the DE plotting techniques of the Section 6.3 and 6.4 applications to the investigation of mechanical systems that exhibit the phenomenon of period-doubling as a selected system parameter is varied.

The Forced Duffing Equation

Section 6.4 in the text introduces the second-order differential equation

$$mx'' + cx' + kx + \beta x^3 = 0 \tag{1}$$

to model the free velocity-damped vibrations of a mass m on a nonlinear spring. Recall that the term kx in Eq. (1) represents the force exerted on the mass by a *linear* spring, whereas the term βx^3 represents the nonlinearity of an actual spring. We want now to discuss the *forced vibrations* that result when an external force $F(t) = F_0 \cos \omega t$ acts on the mass. With such a force adjoined to the system in Eq. (1), we obtain the **forced Duffing equation**

$$mx'' + cx' + kx + \beta x^3 = F_0 \cos \omega t \tag{2}$$

for the displacement $x(t)$ of the mass from its equilibrium position.

If $\beta = 0$ in (2) then we have a linear equation with stable periodic solutions. To illustrate the quite different behavior of a nonlinear system, we take $k = -1$ and $m = c = \beta = \omega = 1$ in Eq. (2), so the differential equation is

$$x'' + x' - x + x^3 = F_0 \cos t \tag{3}$$

As an exercise you may verify that the two critical points $(-1, 0)$ and $(1, 0)$ are stable. We want to examine the dependence of the (presumably steady periodic) response $x(t)$ upon the amplitude F_0 of the periodic external force of period 2π.

First verify that the values $F_0 = 0.60$ and $F_0 = 0.70$ yield the two figures shown on the next page, which indicate a simple oscillation about a critical point if $F_0 = 0.60$, and an oscillation with "doubled period" if $F_0 = 0.70$. In each case the equation was solved numerically with initial conditions $x(0) = 1$, $x'(0) = 0$ and the resulting solution plotted for the range $100 \le t \le 200$ (to show the steady periodic response remaining after the initial transient response has died out). Use *tx*-plots (as in the predator-prey investigation of the 6.3 application) to verify that the period of the oscillation with

$F_0 = 0.70$ is, indeed, twice the period with $F_0 = 0.60$. Then plot analogous figures with $F_0 = 0.75$ and with $F_0 = 0.80$ to illustrate successive *period-doubling* and finally *chaos* as the amplitude of the external force is increased in the range from $F_0 = 0.6$ to $F_0 = 0.8$. This *period-doubling toward chaos* is a common characteristic of the behavior of a *nonlinear* mechanical system as an appropriate physical parameter (such as m, c, k, β, F_0, or ω) in Eq. (2) is increased or decreased. No such phenomenon occurs in linear systems.

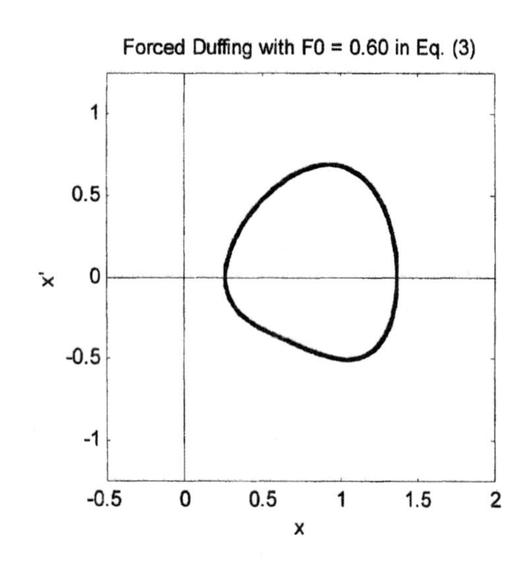

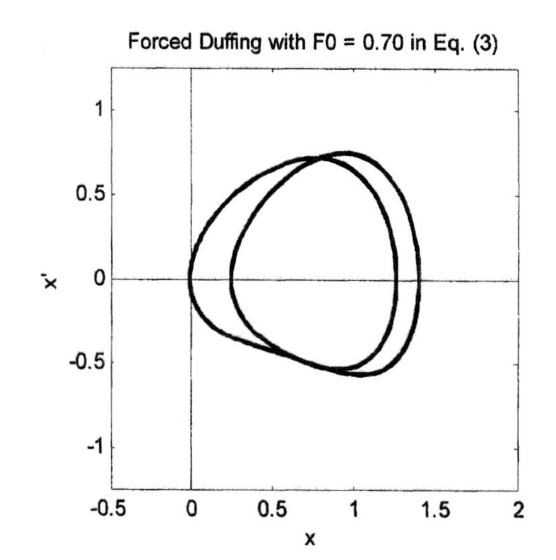

Then investigate the parameter range $1.00 \le F_0 \le 1.10$ for the force constant in Eq. (3). With $F_0 = 1.00$ you should see a period 6π phase plane trajectory that encircles *both* stable critical points (as well as the unstable one). The period doubles around $F_0 = 1.07$ and chaos sets in around $F_0 = 1.10$. See whether you can spot a second period-doubling somewhere between $F_0 = 1.07$ and $F_0 = 1.10$. Produce both phase plane trajectories and *tx*-solution curves on which you can measure the periods.

The Lorenz Strange Attractor

The genesis of the famous 3-dimensional *Lorenz system*

$$\begin{aligned} x'(t) &= -sx + sy \\ y'(t) &= -xz + rx - y \qquad (4) \\ z'(t) &= xy - bz \end{aligned}$$

of differential equations is discussed in the text. A solution curve in *xyz*-space is best visualized by looking at its projection into some *plane*, typically one of the three coordinate planes. The figure on the next page shows the projection into the *xz*-plane of the solution obtained by numerical integration from $t = 0$ to $t = 30$ with the parameter

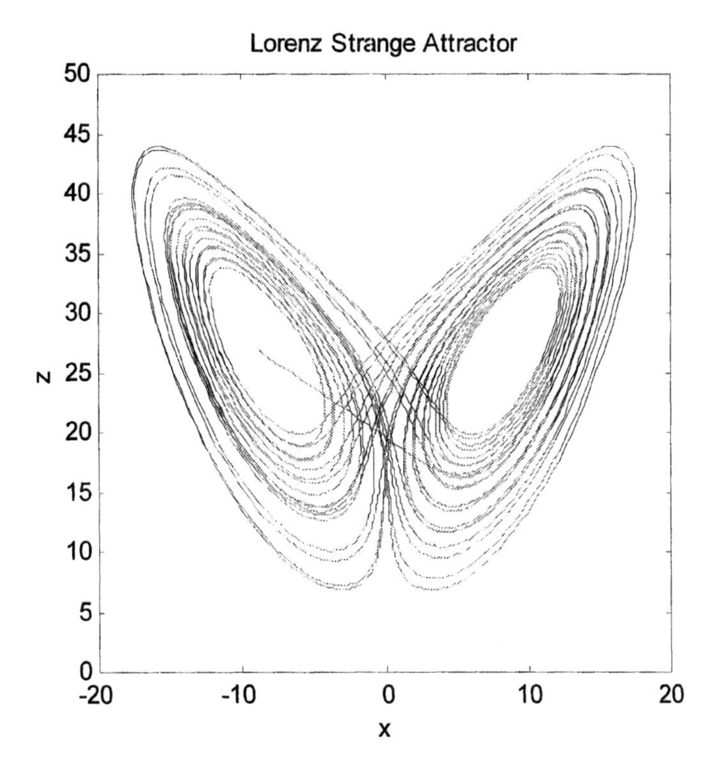

values $b = 8/3$, $s = 10$, $r = 28$ and the initial values $x(0) = -8$, $y(0) = 8$, $z(0) = 27$. As the projection in this figure is traced in "real time", the moving solution point $P(x(t), y(t), z(t))$ appears to undergo a random number of oscillations on the right followed by a random number of oscillations on the left, then a random number of the right followed by a random number on the left, and so on.

A close examination of such projections of the Lorenz trajectory shows that it is *not* simply oscillating back and forth around a pair of critical points (as the figure may initially suggest). Instead, as $t \to \infty$, the solution point $P(t)$ on the trajectory wanders back in forth in space approaching closer and closer to a certain complicated set of points whose detailed structure is not yet fully understood. This elusive set that appears somehow to "attract" the solution point is the famous *Lorenz strange attractor*.

First use an ODE plotting utility to reproduce the *xz*-projection of the Lorenz trajectory shown above. Use the parameter values and initial conditions listed above and numerically integrate the Lorenz system on the interval $0 \le t \le 30$. Plot also the *xy*- and *yz*-projections of this same solution. Next, experiment with different parameter values and initial conditions. For instance, see if you can find a periodic solution with $r = 70$ (and $b = 8/3$, $s = 10$ as before) and initial values $x_0 = -4$ and $z_0 = 64$. To get a trajectory that almost repeats itself, you will need to try different values of y_0 in the range $0 < y_0 < 10$ and look at *xz*-projections.

The Rössler Band

Another much-studied nonlinear three-dimensional system is the Rossler system

$$\begin{aligned} x'(t) &= -y-z \\ y'(t) &= x+ay \\ z'(t) &= b+z(x-c) \end{aligned} \qquad (5)$$

The figure below shows an xy-projection of the *Rössler band*, a chaotic attractor obtained with the values $a = 0.398$, $b = 2$, and $c = 4$ of the parameters in (5). In the xy-plane the Rössler band looks "folded," but in space it appears twisted like a Möbius strip. Investigate the period-doubling toward chaos that occurs with the Rössler system as the parameter a is increased, beginning with $a = 0.3$, $a = 0.35$, and $a = 0.375$ (take $b = 2$ and $c = 4$ in all cases).

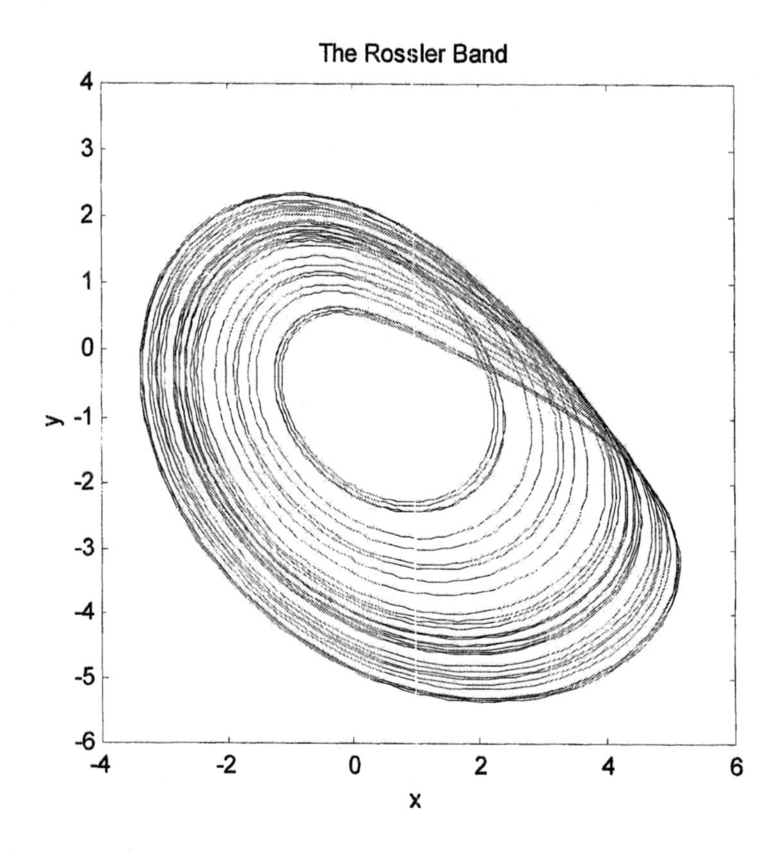

In the following paragraphs we illustrate ODE plotting techniques by showing how to use *Maple*, *Mathematica*, and MATLAB to plot the forced Duffing, Lorenz, and Rössler trajectories pictured here.

Using *Maple*

To plot a periodic trajectory for the forced Duffing equation (3) we need only define the system

```
F0 := 0.60:
m:=1:    c:=1:     k:=-1:     b:=1:     w:=1:
deq1 :=  diff(x(t),t) = y:
deq2 :=
m*diff(y(t),t) = -c*y - k*x - b*x^3 + F0*cos(w*t):
```

and then load the **DEtools** package and use the **DEplot** function. For instance, the command

```
with(DEtools):
DEplot([deq1,deq2], [x,y],
       t=100..200, {[x(0)=1, y(0)=0]},
       stepsize=0.1,
       linecolor=blue, arrows=none);
```

plots the periodic trajectory obtained with $F_0 = 0.60$.

To picture the Rössler band, we first define the Rössler system

```
a := 0.398:
b := 2:   c := 4:
deq1 :=   diff(x(t),t) = -y - z:
deq2 :=   diff(y(t),t) = x + a*y:
deq3 :=   diff(z(t),t) = b + z*(x - c):
```

and then plot the desired *xy*-projection:

```
with(DEtools):
DEplot([deq1,deq2,deq3], [x,y,z],
       t=0..200, {[ x(0)=1,y(0)=1,z(0)=1 ]},
       stepsize=0.1,
       scene = [x,y], thickness=0,
       linecolor=blue, arrows=none);
```

Using *Mathematica*

To plot a periodic trajectory for the forced Duffing equation (3) we need only define the system

```
F0 = 0.60;
m = 1;   c = 1;    k = -1;    b = 1;    w = 1;
```

```
deq1 =    x'[t] == y[t];
deq2 =
m y'[t] == -c y[t] - k x[t] - b x[t]^3 + F0 Cos[w t];
```

and then use **NDSolve** to integrate numerically. For instance, the command

```
soln = NDSolve[ {deq1, deq2,
                  x[0]==1, y[0]==0},
                 {x[t],y[t]}, {t,0,200},
                  MaxSteps->2500 ];;
```

yields an approximate solution on the interval $0 \le t \le 200$ satisfying the initial conditions $x(0)=1,\ y(0)=0$. Then the command

```
ParametricPlot[
     Evaluate[{x[t],y[t]} /. soln], {t,100,200}];
```

plots the periodic trajectory corresponding to $F_0 = 0.60$.

To picture the Rössler band, we first define the Rössler system

```
a = 0.398;
b = 2;    c = 4;
deq1 =    x'[t] == -y[t] - z[t];
deq2 =    y'[t] == x[t] + a y[t];
deq3 =    z'[t] == b + z[t] (x[t] - c);
```

solve numerically,

```
soln =
NDSolve[ {deq1, deq2, deq3,
           x[0]==1, y[0]==1, z[0]==1},
          {x[t],y[t],z[t]}, {t,0,200},
           MaxSteps->3000];
```

and then plot the *xy*-projection:

```
ParametricPlot[
   Evaluate[{x[t],y[t]} /. soln], {t,0,200}];
```

Using MATLAB

To plot a periodic trajectory for the forced Duffing equation (3) we first define the MATLAB function

```
function  yp = duffing(t,y)
F0 = 0.60;
m = 1;   c = 1;    k = -1;    b = 1;    w = 1;
yp = y;
x = y(1);
y = y(2);
yp(1)  = y;
yp(2)  =  (-c*y  -  k*x  -  b*x.^3  +  F0*cos(t))/m;
```

as the m-file **duffing.m** defining the forced Duffing system. Note that it specifies the desired numerical values of the parameters in Eq. (2). Then the commands

```
[t,y]  = ode45('duffing',0:0.1:100,[1;0]);
n = length(t);
[t,y]  = ode45('duffing',100:0.1:200,y(n,:));
plot(y(:,1),y(:,2));
```

plot the periodic xy-trajectory corresponding to $F_0 = 0.60$. Note that we solve first from $t = 0$ to $t = 100$, then use the final value here as the initial value in solving from $t = 100$ to $t = 200$.

To plot the Lorenz strange attractor, we first define the MATLAB function

```
function  yp = lorenz(t,y)
yp = y;
b = 8/3;   s = 10;   r = 28;
x = y(1);
z = y(3);
y = y(2);
yp(1)  = -s*x + s*y;
yp(2)  = -x.*z + r*x - y;
yp(3)  =   x.*y  - b*z;
```

as the m-file **lorenz.m**. Then we solve numerically

```
[t,y]  = ode45('lorenz',0:0.005:30,[-9;8;27]);
```

and finally plot the *xz*-projection:

```
plot(y(:,1),y(:,3))
```

Further Investigation

We noted above that the Lorenz trajectory is not simply oscillating around a pair of critical points, despite what a static diagram may suggest. Using *Mathematica*, we can animate the trajectory in three dimensions so as to get a "real-time" view of the wandering solution point $P(t)$. (*Maple* and MATLAB users can substitute equivalent commands.)

The following code will animate the trajectory of the Lorenz system (4) over $0 \le t \le 30$, using as before $\beta = 8/3$, $\sigma = 10$, $\rho = 28$, $x(0) = -8$, $y(0) = 8$, and $z(0) = 27$. First we define the system...

```
deq1 = x'[t] == -10x[t] + 10y[t];
deq2 = y'[t] == -x[t]*z[t] + 28x[t] - y[t];
deq3 = z'[t] == x[t]*y[t] - (8/3)z[t];
```

...solve it numerically for the stated initial conditions...

```
soln = First[NDSolve[{deq1, deq2, deq3,
                x[0] == 8, y[0] == 8, z[0] == 27},
                {x, y, z}, {t, 0, 30},
                MaxSteps -> Infinity]];
```

...load in a needed *Mathematica* package...

```
<<Graphics`Animation`
```

...and finally use the following `Do` loop to generate the frames of the animation:

```
Do[ endpoint = Graphics3D[{PointSize[0.025], Hue[0],
          Point[{x[t], y[t], z[t]}/.soln/.t -> s]}];
    curve = ParametricPlot3D[
               Evaluate[{x[t],y[t], z[t]}/.soln],
               {t, 0, s}, PlotPoints -> 1500,
               DisplayFunction ->Identity];
    lorenz = Show[endpoint, curve,
               BoxRatios -> {1, 1, 1},
```

```
        PlotRange -> {{-20,20}, {-25,25}, {0,50}},
        ImageSize -> 6*72,
        ViewPoint -> {3.712, 0.000, 0.795},
        DisplayFunction -> $DisplayFunction],
    {s, 0.125, 30, 0.125}];
```

(Double-click on any of the frames to play the animation; during play you can use the number keys 1,2,3,... to adjust the speed of play.)

Do you see any pattern in the wanderings of the red dot? Were we correct in saying that it does not simply oscillate left-right-left-right?

It is also interesting to see the Lorenz attractor from all sides. The following command will spin around the final frame of the preceding animation; play the resulting movie as indicated above:

```
SpinShow[lorenz]
```

You are also invited to modify the above commands so as to draw the Lorenz attractor using different starting points and/or different values of β, σ, and ρ.

Chapter 7

Laplace Transform Methods

Application 7.1
Computer Algebra Transforms and Inverse Transforms

If $f(t) = t\cos(3t)$ then the definition of the Laplace transform gives the improper integral

$$F(s) = \mathcal{L}\{f(t)\} = \int_0^\infty t\,e^{-st}\cos 3t\,dt \qquad (1)$$

whose evaluation would appear to require a fairly tedious integration by parts. Consequently, a computer algebra system is useful for the quick calculation of Laplace transforms.

In the paragraphs that follow we illustrate the use of appropriate *Maple*, *Mathematica*, and MATLAB commands to find Laplace transforms and inverse transforms. You can use these computer algebra commands to check the answers to Problems 11–32 in Section 7.1 of the text, as well as a few interesting problems of your own selection.

Using *Maple*

First we load *Maple*'s **int**egral **trans**forms package **inttrans** with the command

```
with(inttrans):
```

and define (as an expression) the function $f(t)$ whose transform appears in (1):

```
f := t*cos(3*t):
```

Then the Laplace transform of f is given by

```
F := laplace(f, t, s);
```

$$F := \frac{\cos\left(2\arctan\left(3\frac{1}{s}\right)\right)}{s^2+9}$$

```
F := simplify(expand( F ));
```

$$F := \frac{s^2-9}{\left(s^2+9\right)^2}$$

Thus we obtain the Laplace transform $F(s) = (s^2-9)/(s^2+9)^2$.

As illustrated above, it's a good idea to routinely "expand and simplify" the result when calculating a Laplace transform. We can inverse Laplace transform to recover the original function $f(t)$ with the *Maple* command

```
invlaplace(F, s, t);
```

$$t\cos(3t)$$

Thus we are (as desired) back where we started.

Remark Note carefully the order of s and t in the commands above — first t, then s when transforming; first s, then t when inverse transforming.

Using *Mathematica*

First we load the Laplace transforms package **Calculus:LaplaceTransform** with the command

```
Needs["Calculus`LaplaceTransform`"]
```

and define (as an expression) the function $f(t)$ whose transform appears in (1):

```
f = t Cos[3 t];
```

Then the Laplace transform of f is given by

```
F = LaplaceTransform[f, t, s]
```

$$\frac{2s^2}{(9+s^2)^2} - \frac{1}{9+s^2}$$

```
F = F // Expand // Simplify
```

$$\frac{-9+s^2}{(9+s^2)^2}$$

Thus we obtain the Laplace transform $F(s) = (s^2-9)/(s^2+9)^2$.

As illustrated above, it's a good idea to routinely "expand and simplify" the result when calculating a Laplace transform. We can inverse Laplace transform to recover the original function $f(t)$ with the *Mathematica* command

```
InverseLaplaceTransform[F, s, t] // Expand // Simplify
```

$t\cos(3t)$

Thus we are (as desired) back where we started.

Remark Note carefully the order of s and t in the commands above — first t, then s when transforming; first s, then t when inverse transforming.

Using MATLAB

First we define (as an expression) the function $f(t)$ whose transform appears in (1):

```
syms t s
f = t*cos(3*t);
```

Then the Laplace transform of f is given by

```
F = laplace(f)
F =
1/(s^2+9)*cos(2*atan(3/s))

F = simplify(expand(F))
F =
(s^2-9)/(s^2+9)^2
```

Thus we obtain the Laplace transform $F(s) = (s^2-9)/(s^2+9)^2$.

As illustrated above, it's a good idea to routinely "expand and simplify" the result when calculating a Laplace transform. We can inverse Laplace transform to recover the original function $f(t)$ with the MATLAB command

```
ilaplace(F)
ans =
t*cos(3*t)
```

Thus we are (as desired) back where we started.

Remark Note that the Laplace transform of a function of t is automatically a function of the variable s, while the inverse transform of a function of s is automatically a function of the variable t. So if we adhere to the usual notations, then variables need not be specified in calculating transforms and inverse transforms with MATLAB.

Application 7.2

Transforms of Initial Value Problems

The typical computer algebra system knows Theorem 1 and its Corollary in Section 7.2 of the text, and hence can transform not only functions (as in the Section 7.1 application), but whole initial value problems. Here we illustrate this facility by applying *Maple*, *Mathematica*, and MATLAB to solve the initial value problem

$$x'' + 4x = \sin 3t, \qquad x(0) = x'(0) = 0 \tag{1}$$

of Example 2 in the text. You can try it for yourself with the initial value problems in Problems 1 through 16 there.

Using *Maple*

First we load the Laplace transforms package **inttrans** with the command

```
with(inttrans):
```

and define the differential equation and initial conditions that appear in (1):

```
de :=  diff(x(t),t$2)+4*x(t)=sin(3*t):
inits :=  {x(0)=0, D(x)(0)=0}:
```

The Laplace transform of the differential equation is given by

```
DE := laplace(de, t, s);
```

$$\mathrm{DE} := s\,(s\,\mathrm{laplace}(x(t),t,s) - x(0)) - D(x)(0) + 4\,\mathrm{laplace}(x(t),t,s) = 3\frac{1}{s^2+9}$$

The result of this command is a linear (algebraic) equation in the as yet unknown transform $\mathrm{laplace}(x(t),t,s)$. We proceed to solve for this transform $X(s)$ of the unknown function $x(t)$,

```
X(s) := solve(DE, laplace(x(t), t, s));
```

$$X(s) := \frac{x(0)\,s^3 + 9\,s\,x(0) + D(x)(0)\,s^2 + 9\,D(x)(0) + 3}{s^4 + 13s^2 + 36}$$

and substitute the initial conditions,

```
X(s) := subs(inits, X(s));
```

$$X(s) := 3\frac{1}{s^4+13s^2+36}$$

Finally we need only inverse transform to find the solution $x(t)$ of the initial value problem in (1).

```
x(t) := invlaplace(X(s), s, t);
```

$$x(t) := \frac{3}{10}\sin(2t) - \frac{1}{5}\sin(3t)$$

Of course we could probably get this result immediately with **dsolve**, but the intermediate output generated by the steps above can be quite instructive.

Using *Mathematica*

First we load the Laplace transforms package **Calculus:LaplaceTransform** with the command

```
Needs["Calculus`LaplaceTransform`"]
```

and define the differential equation and initial conditions that appear in (1):

```
de =     x''[t] + 4 x[t] == Sin[3t];
inits = {x[0]->0,   x'[0]-->0};
```

Then the Laplace transform of the differential equation is given by

```
DE = LaplaceTransform[de, t, s]
```

$$\text{LaplaceTransform}(x(t),t,s)\, s^2 - x(0)\, s + 4\, \text{LaplaceTransform}(x(t),t,s) - x'(0) == \frac{3}{s^2+9}$$

The result of this command is a linear (algebraic) equation in the as yet unknown $\text{LaplaceTransform}(x(t),t,s)$. We proceed to solve for this transform $X(s)$ of the unknown function $x(t)$,

```
X = Solve[DE, LaplaceTransform[x[t], t, s]]
```

$$\{\{\text{LaplaceTransform}(x(t),t,s) \to -\frac{-s\,x(0) - x'(0) - \frac{3}{s^2+9}}{s^2+4}\}\}$$

```
X = X // Last // Last // Last
```

$$-\frac{-s\,x(0)-x'(0)-\dfrac{3}{s^2+9}}{s^2+4}$$

and substitute the initial conditions,

```
X =  X /. inits
```

$$\frac{3}{(s^2+4)(s^2+9)}$$

Finally we need only inverse transform to find $x(t)$.

```
x = InverseLaplaceTransform[X, s, t] // Expand
```

$$\frac{3}{10}\sin(2t)-\frac{1}{5}\sin(3t)$$

Of course we could probably get this result immediately with **DSolve**, but the intermediate output generated by the steps above can be quite instructive.

Using MATLAB

To use MATLAB it is convenient to rewrite equation (1) in the form

$$x''+4x-\sin 3t = 0 .$$

Then we enter the symbolic expression on the left-hand side to define our differential equation:

```
syms s t x X
de = diff(sym('x(t)'),t,2)+4*sym('x(t)')-sin(3*t)

de =
diff(x(t),`$`(t,2)) + 4*x(t) - sin(3*t)
```

The Laplace transform of the differential equation is given by

```
DE = laplace(de)
DE =
s*(s*laplace(x(t),t,s) - x(0)) - D(x)(0) +
          4*laplace(x(t),t,s) - 3/(s^2+9)
```

At this point we substitute the initial conditions $x(0) = x'(0) = 0$,

```
DE = subs(DE,{'x(0)','D(x)(0)'},{0,0})
```

```
DE =
s^2*laplace(x(t),t,s)+4*laplace(x(t),t,s)-3/(s^2+9)
```

The result (understood to be equated to 0) is a linear (algebraic) equation in the as yet unknown transform **`X = laplace(x(t),t,s)`**. We proceed to solve for this transform **`X`** of the unknown function **`x(t)`**,

```
DE = subs(DE,'laplace(x(t),t,s)','X')

DE =
s^2*X+4*X-3/(s^2+9)

X = solve(DE,X)

X =
3/(s^2+9)/(s^2+4)
```

Finally we need only find the inverse Laplace transform of $X(s)$ to obtain the solution $x(t)$ of the initial value problem in (1).

```
x = ilaplace(X,s,t)

x =
-1/5*sin(3*t)+3/10*sin(2*t)

pretty(x)

- 1/5 sin(3 t) + 3/10 sin(2 t)
```

Of course we could probably get this result immediately with **`dsolve`**, but the intermediate output generated by the steps above can be quite instructive.

Further Investigation

Now that we are familiar with finding Laplace transforms using a computer algebra system, let's pursue the above example $f(t) = t\cos 3t$ (as well as several others) from a different angle, asking this question:

> For $n = 1,2,3,\ldots$ suppose we form the function $f_n(t) = t\cos 3t \sin nt$, the product of $f(t)$ and $\sin nt$; *how will the Laplace transform* $\mathcal{L}\{f_n(t)\}$ *behave as n grows?*

Before throwing up your hands and declaring that you have no idea, think it over a moment. If you knew the graph of $f(t)$, then what would the graph of $f_{1000}(t)$ look like? And how would this affect the Laplace transform?

The following commands will calculate $\mathcal{L}\{f_n(t)\}$ in *Mathematica*; *Maple* and MATLAB users can substitute equivalent commands. (Without the **Assumptions** option below you would wait a lot longer while *Mathematica* figures out what to do with if $n = -3$, for example.)

```
fn := t*Cos[t]*Sin[n*t]
Fn = LaplaceTransform[fn, t, s,
          Assumptions -> (n > 0)]
```

Then the following commands will generate movie frames showing the graphs of $f_n(t)$ on the left and of $\mathcal{L}\{f_n(t)\}$ on the right for $n = 1,2,\ldots,25$. Double-click on any of the frames to play the animation; during play you can use the number keys 1,2,3,... to adjust the speed of play:

```
fplot[n_] := Plot[f[t, n], {t, 0, 5},
   PlotRange -> {-4,4}, DisplayFunction -> Identity]

Fplot[n_] := Plot[F[s, n], {s, 0, 5},
   PlotRange -> {0,0.4}, DisplayFunction -> Identity]

Table[Show[GraphicsArray[{fplot[n],Fplot[n]}],
   ImageSize -> 9*72], {n, 1, 25}];
```

How does the Laplace transform of $f_n(t)$ behave as *n* grows? Does this confirm your conjecture (or perhaps lead you to form one now)?

Let's try it again, but this time using $f_n(t) = t\cos 3t \cos nt$ instead. Modify the above commands as needed and play the animation; does $\mathcal{L}\{f_n(t)\}$ still seem to drop off to zero for all *s* as *n* grows? (You may want to change the viewing windows in the **Plot** commands to help you see things more clearly.)

Is something going on here? Let's investigate further by starting with each of the following choices for $f(t)$, and then multiplying by $\sin nt$ and/or $\cos nt$ to form $f_n(t)$. Then once again modify the *Mathematica* commands above to generate an animation, taking note of how $\mathcal{L}\{f_n(t)\}$ behaves as *n* increases:

- $f(t) = e^t$ (written **Exp[t]** in *Mathematica*)
- $f(t) = \sqrt{t}$ (written **Sqrt[t]**)
- $f(t) = t^2 - 3t$

Now that you're pretty convinced that what you're seeing is no coincidence, can you use the definition of Laplace transform to explain this behavior? (Remember that $\mathcal{L}\{f_n(t)\} = \int_0^\infty e^{-st} f_n(t)\, dt$ can be interpreted as the *total signed area* under the graph of $e^{-st} f_n(t)$ over $0 \le t < \infty$; what happens to this area as n grows?)

Finally, just to stir the pot once more, use *Mathematica* as above to investigate the case $f(t) = 1/t$; does $\mathcal{L}\{f_n(t)\}$ still drop off as n grows? Why not? Does your conjecture need some refining?

Project 7.3

Damping and Resonance Investigations

Here we outline *Maple*, *Mathematica*, and MATLAB investigations of the behavior of the mass-spring-dashpot system

$$mx'' + cx' + kx = F(t), \quad x(0) = x'(0) = 0 \tag{1}$$

with parameter values $m = 25$, $c = 10$, and $k = 226$ in response to a variety of possible external forces:

1. $F(t) \equiv 226$

This constant force should give damped oscillations "leveling off" to a constant solution (why?).

2. $F(t) \equiv 900\cos 3t$

With this periodic external force you should see a steady periodic oscillation with an exponentially damped transient motion (as illustrated in Fig. 3.6.13 in the text).

3. $F(t) \equiv 900e^{-t/5}\cos 3t$

Now the periodic external force is exponentially damped, and the transform $X(s)$ involves a repeated quadratic factor that signals the presence of a resonance phenomenon. The response $x(t)$ is a constant multiple of that shown in Fig. 7.3.5 in the text.

4. $F(t) \equiv 900te^{-t/5}\cos 3t$

We have inserted a t-factor to make it a bit more interesting. The response $x(t)$ is plotted in Fig. 7.3.6 in the text.

5. $F(t) \equiv 16200t^3e^{-t/5}\cos 3t$

Now you'll find that the transform $X(s)$ involves the *fifth* power of a quadratic factor, and its inverse transform by manual methods would be impossibly tedious.

To see the advantage of using Laplace transforms, you might set up the appropriate differential equation **de** for case 5 and take a look at the result of the commands

```
dsolve({de, x(0)=0,D(x)(0)=0}, x(t));        (Maple)
```

```
DSolve[{de, x[0]==0,x'[0]==0},x[t],t]          (Mathematica)

dsolve(de, 'x(0)=0, Dx(0)=0')                 (MATLAB)
```

Of course you can substitute you own favorite mass-spring-dashpot parameters for those used above. However, it will simplify the calculations if you choose m, c, and k so that

$$mr^2 + cr + k = (pr + a)^2 + b^2 \tag{2}$$

where p, a, and b are integers. One way is to select the latter integers first, then use (2) to determine m, c, and k.

Using *Maple*

First we define the mass-spring dashpot parameters

```
m := 25:     c := 10:     k := 226:
```

and the external force function

```
F := 900*t*exp(-t/5)*cos(3*t);
```

for case 4. Then our differential equation is defined by

```
de := m*diff(x(t),t$2) + c*diff(x(t),t) + k*x(t) = F;
```

and the initial conditions are given by

```
inits := {x(0)=0, D(x)(0)=0}:
```

Now we apply the Laplace transform to this equation, solve for the transform $X(s)$ of $x(t)$, and substitute the initial conditions.

```
with(inttrans):

DE := laplace(de, t,s);

X(s) := solve(DE, laplace(x(t), t,s));

X(s) := subs(inits, X(s));
```

At this point the command `factor(denom(X))` shows that

$$X(s) = \frac{22500(25s^2 + 10s - 224)}{(25s^2 + 10s + 226)^3}.$$

The cubed quadratic factor would be difficult to handle manually, but Maple readily calculates the inverse transform

```
x(t) := invlaplace(X(s) ,s,t);
```

$$x(t) := 3e^{\left(-\frac{1}{5}t\right)} t^2 \sin(3t) + t\,e^{\left(-\frac{1}{5}t\right)} \cos(3t) - \frac{1}{3} e^{\left(-\frac{1}{5}t\right)} \sin(3t)$$

Let's collect the coefficients

```
A := t:
B := 3*t^2-1/3:
```

Then our solution has the form

$$x(t) = C(t)\cos(3t - \alpha)$$

where the time-varying amplitude function for these damped oscillations is defined by

```
C(t) := sqrt(A^2 + B^2)*exp(-t/5);
```

$$C(t) := \frac{1}{3}\sqrt{-9t^2 + 81t^4 + 1}\, e^{\left(-\frac{1}{5}t\right)}$$

Finally, the command

```
plot([x(t), C(t), -C(t)], t=0..40);
```

produces the plot shown below. The resonance resulting (in effect) from the repeated quadratic factor visible in the Laplace transform of the solution $x(t)$ consists of a temporary buildup before the oscillations are damped out.

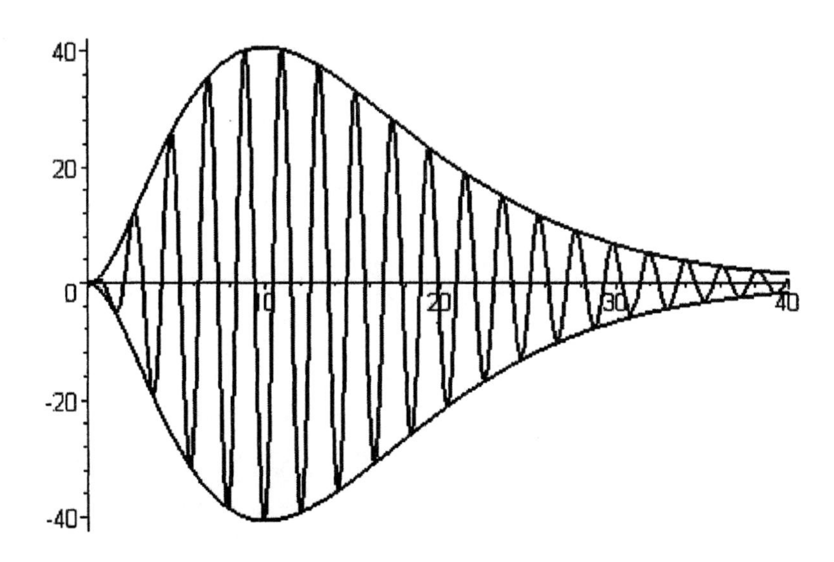

Using *Mathematica*

First we define the mass-spring dashpot parameters

```
m = 25;     c = 10;     k = 226;
```

and the external force function

```
F = 900 Exp[-t/5] Cos[3t];
```

for case 3. Then our differential equation is defined by

```
de =   m x''[t] + c x'[t] + k x[t] == F
```

and the initial conditions are given by

```
inits = {x[0]->0,  x'[0]->0};
```

Now we apply the Laplace transform to this equation, solve for the transform $X(s)$ of $x(t)$, and substitute the initial conditions.

```
Needs["Calculus`LaplaceTransform`"]
DE = LaplaceTransform[ de, t, s ]
X =  Solve[DE, LaplaceTransform[x[t],t,s]]
X =  X // Last // Last // Last
X =  X /. inits // Simplify
```

$$\frac{4500(5s+1)}{(25s^2+10s+226)^2}$$

The repeated quadratic factor visible here in the denominator of the Laplace transform $X(s)$ signals a resonance phenomenon. We now inverse transform to get the solution

```
x = InverseLaplaceTransform[X, s, t] // Expand
```

$$6e^{-t/5}t\sin(3t)$$

Thus we have a damped oscillation with the time-varying amplitude function

```
c = 6 t Exp[-t/5];
```

When we plot the solution curve and envelope curves,

```
Plot[ {x, c, -c}, {t,0,30}];
```

we see that the "resonance" consists in the buildup in the amplitude of the forced oscillations before the damping prevails.

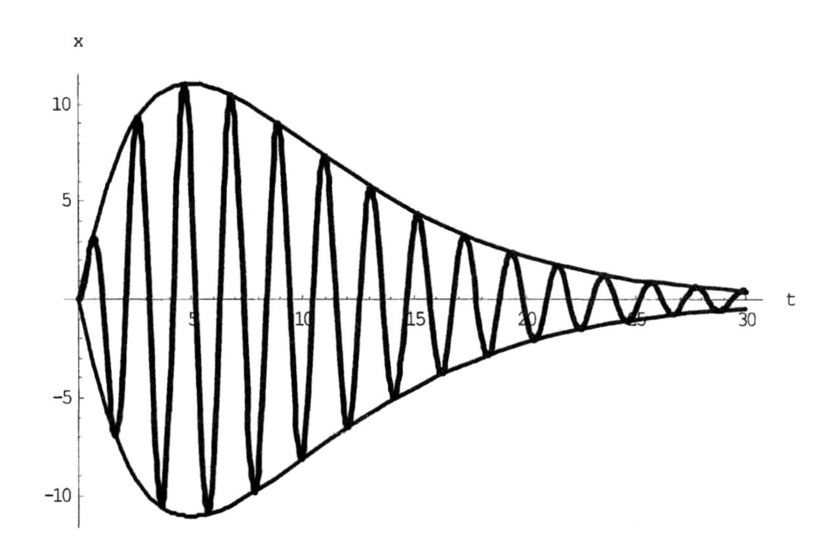

Using MATLAB

First we define the mass-spring dashpot parameters

```
m = 25;        c = 10;        k = 226;
```

and the differential equation

```
syms s t x X
x = sym('x(t)');
de = m*diff(x,t,2) + c*diff(x,t) +k*x -
                    16200*t^3*exp(-t/5)*cos(3*t)
```

corresponding to case 5.

Now we apply the Laplace transform to this equation, substitute the initial conditions $x(0) = x'(0) = 0$, and solve for the transform $X(s)$ of $x(t)$.

```
DE = laplace(de);
DE = subs(DE,{'x(0)','D(x)(0)'},{0,0});
DE = subs(DE,'laplace(x(t),t,s)','X')
```

```
X = solve(DE,X);

pretty(X)

                        2          4          3
           -33600 s  + 625 s  + 500 s  - 13480 s + 49276
60750000 ---------------------------------------------
                                      2          5
                          (226 + 25 s  + 10 s)
```

The quintic quadratic factor visible in the denominator of the Laplace transform here would be pretty discouraging if we were working manually, but MATLAB readily finds the inverse transform

```
x = ilaplace(X,s,t)
x =
27*exp(-1/5*t)*t^4*sin(3*t) + 18*t^3*exp(-1/5*t)*cos(3*t)
- 9*exp(-1/5*t)*t^2*sin(3*t) - 3*exp(-1/5*t)*t*cos(3*t)
+ exp(-1/5*t)*sin(3*t)
```

where we see the coefficients

$$A = 18t^3 - 3 \quad \text{and} \quad B = 27t^4 - 9t^2 + 1$$

of the damped $\cos(3t)$ and $\sin(3t)$ terms, respectively. Then our solution has the damped oscillatory form

$$x(t) = C(t)\exp(-t/5)\cos(3t - \alpha)$$

where the time-varying coefficient $C(t)$ is defined by $C = \sqrt{A^2 + B^2}$.

We can therefore proceed to plot this damped oscillation with the commands

```
ezplot(x,0,50)

axis([0 50 -10^5 10^5])

hold on

t = 0 : 0.2 : 60;

A = 18*t.^3 - 3;

B = 27*t.^4 - 9*t.^2 + 1;

C = sqrt(A.^2 + B.^2);

plot(t,C.*exp(-t/5),'k')

plot(t,-C.*exp(-t/5),'k')
```

which produce the lovely plot shown below. The resonance resulting (in effect) from the repeated quadratic factor visible in the Laplace transform of the solution $x(t)$ consists of a temporary buildup before the oscillations are damped out. Note the exceptional "flatness" of the solution curve at the origin, resulting from the t^3-factor in the external force function, and the consequent high multiplicity of the repeated quadratic.

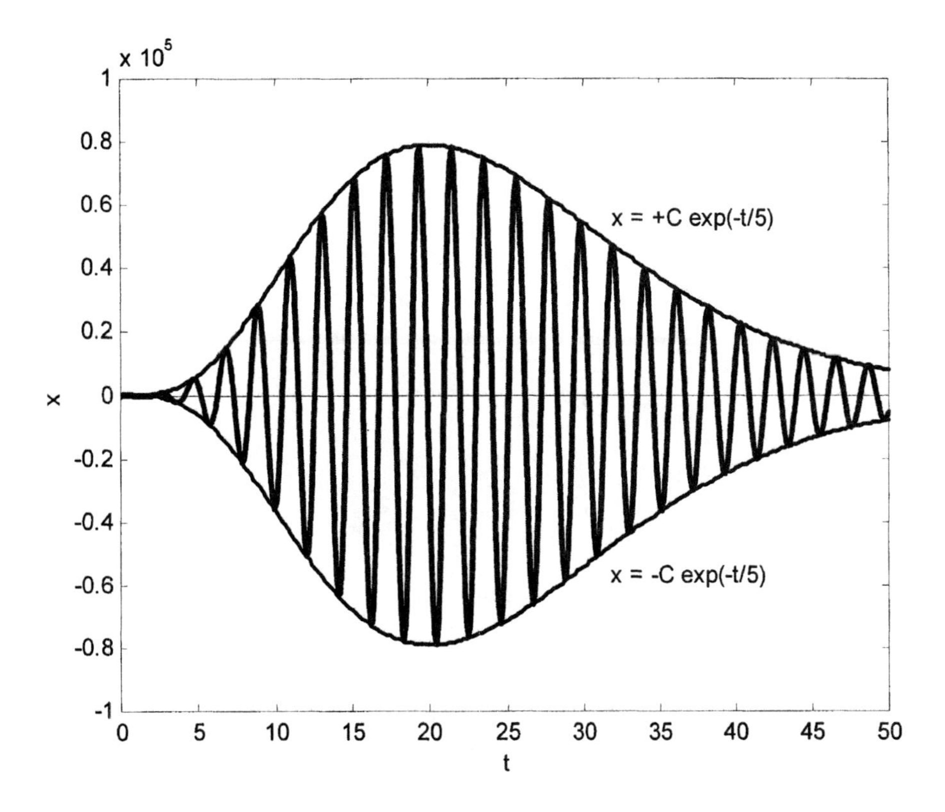

Application 7.5
Engineering Functions

Periodic piecewise linear functions occur so frequently as input functions in engineering applications that they are sometimes called **engineering functions.** Computations with such functions are readily handled by computer algebra systems. In the paragraphs that follow we first show how to define typical engineering functions — such as sawtooth, triangular-wave, and square-wave functions — using *Maple*, *Mathematica*, and MATLAB, and then illustrate the solution of a mass-spring-dashpot problem

$$mx'' + cx' + kx = F(t), \qquad x(0) = x'(0) = 0 \tag{1}$$

when the external force function $F(t)$ is an engineering function. Our main object is to illustrate steady periodic responses to periodic inputs, so it is convenient to use numerical rather than Laplace transform methods. For your own investigation you can solve similarly the initial value problem in (1) with various mass-spring-dashpot parameters — for instance, selected digits of you student ID number — and with input engineering functions having various amplitudes and periods.

Using *Maple*

Typical engineering functions are defined by

```
SawTooth := t -> 2*t - 2*floor(t) - 1:

TriangularWave :=
     t -> abs(2*SawTooth((2*t-1)/4)) - 1:

SquareWave := t -> signum(TriangularWave(t)):
```

A plot of each of these functions verifies that it has period 2 and that its name is aptly chosen. For instance, the result of the command

```
plot(SquareWave(t), t = 0..6);
```

shows (at the top of the next page) three square-wave cycles of length 2.

If $f(t)$ is one of the three period 2 engineering functions defined above, then the function $f(2t/p)$ will have period p. To illustrate this, try

```
plot(TriangularWave(2*t/p), t = 0..3*p);
```

with various values of p. The resulting graph (on the top of the next page) should show three triangular-wave cycles of length p.

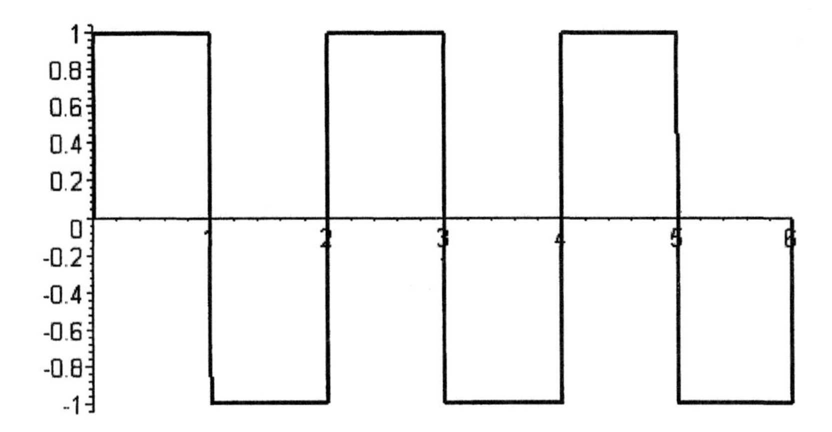

Now let's consider the mass-spring-dashpot equation in (1) with selected parameter values and an input forcing function having period p and amplitude F_0, for instance,

```
m := 4:        c := 8:        k := 5:
p := 1:
F0 := 4:
input := t -> F0*SquareWave(2*t/p);
```

You can plot this **`input`** function to verify that it has period 1:

```
plot(input(t), t = 0..5);
```

Then our differential equation is

```
de :=
m*diff(x(t),t$2) + c*diff(x(t),t) + k*x(t) = input(t):
```

Next, let's suppose that the mass is initially at rest in its equilibrium position and solve numerically the resulting initial value problem.

```
response :=
dsolve( {diffEq, x(0)=0, D(x)(0)=0}, x(t),
        type=numeric);
```

When we plot this solution, we see that after an initial transient dies out, the response function $x(t)$ settles down (as expected?) to a periodic oscillation with the same period as the input.

```
with(plots):
odeplot(response, [t,x(t)], 0..10, numpoints=300);
```

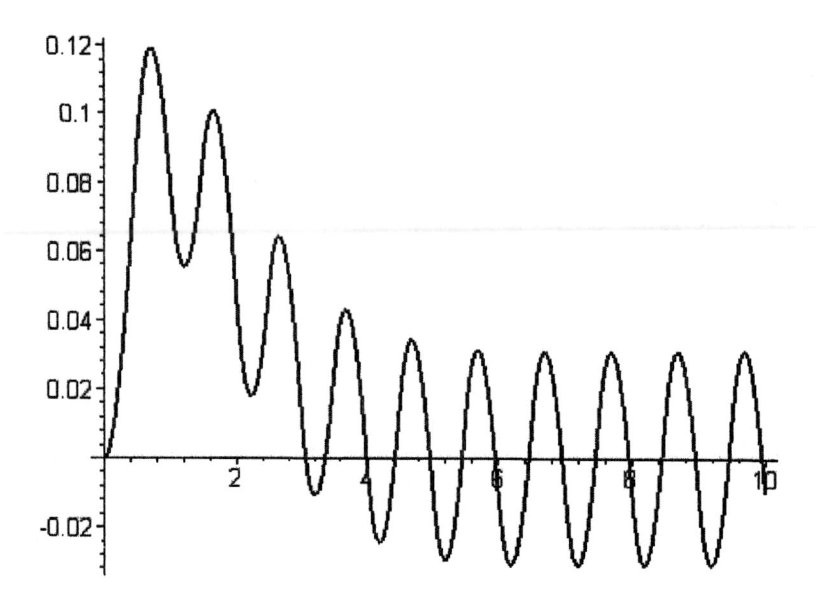

Using *Mathematica*

Typical engineering functions are defined by

```
SawTooth[t_] := 2t - 2 Floor[t] - 1

TriangularWave[t_] := Abs[2 SawTooth[(2t-1)/4]] - 1

SquareWave[t_] := Sign[TriangularWave[t]]
```

A plot of each of these functions verifies that it has period 2 and that its name is aptly chosen. For instance, the result of the command

```
Plot[SquareWave[t], {t,0,6}];
```

shows three square-wave cycles of length 2. (Try it.)

If $f(t)$ is one of the three period 2 engineering functions defined above, then the function $f(2t/p)$ will have period p. To illustrate this, try

```
p = 5;
Plot[TriangularWave[2 t/p], {t, 0, 3p}];
```

with various values of p. The resulting graph should show three triangular-wave cycles of length p. For instance, with $p = 5$ we get the period 5 triangular-wave graph shown at the top of the next page.

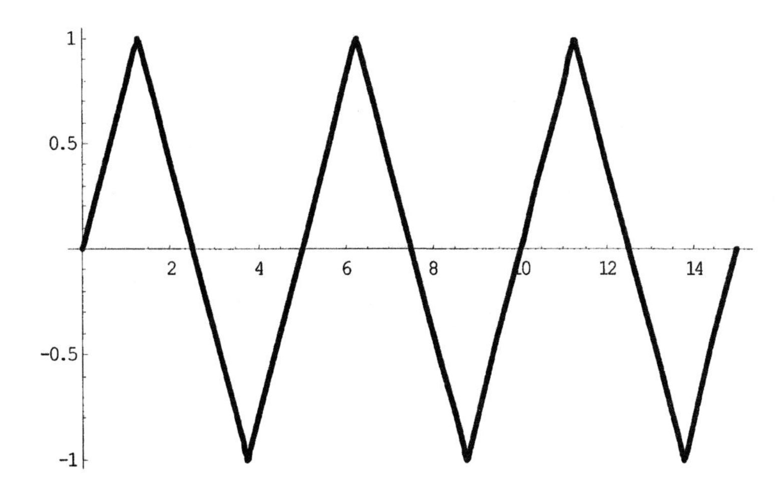

Now let's consider the mass-spring-dashpot equation in (1) with selected parameter values and an input forcing function having period p and amplitude F_0, for instance,

```
m = 3;      c = 35;      k = 10;
p = 2;
F0 = 15;
input = F0 TriangularWave[2t/p];
```

You can plot this **input** function to verify that it has period 2:

```
Plot[ input, {t, 0, 15} ];
```

Then our differential equation is

```
de =   m x''[t] + c x'[t] + k x[t] == input
```

Next, let's suppose that the mass is initially at rest in its equilibrium position and solve numerically the resulting initial value problem.

```
response =
NDSolve[{de, x[0]==0, x'[0]==0}, x, {t,0,15}]
```

When we plot this solution, we see that after an initial transient dies out, the response function $x(t)$ settles down (as expected?) to a periodic oscillation with the same period $p = 2$ as the input.

```
Plot[x[t] /. response, {t,0,10}];
```

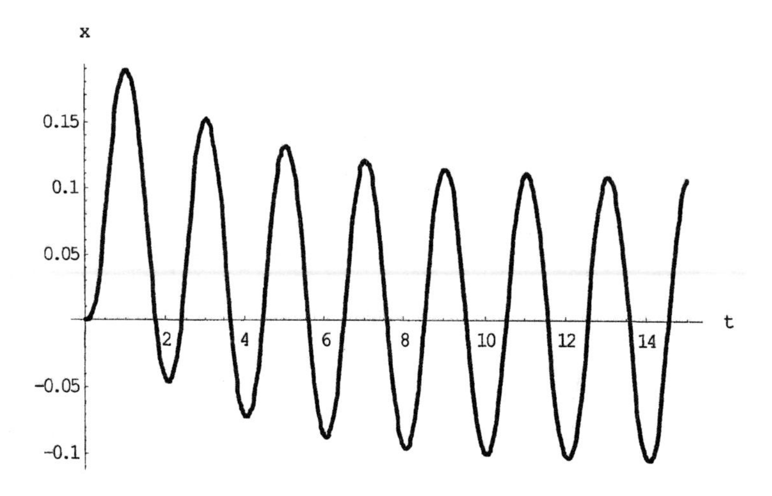

Using MATLAB

Typical engineering functions are defined by

```
function    x = sawtooth(t)
x = 2*t - 2*floor(t) - 1;

function    x = triangle(t)
x = abs(2*sawtooth((2*t-1)/4)) - 1;

function    x = square(t)
x = sign(triangle(t));
```

A plot of each of these functions verifies that it has period 2 and that its name is aptly chosen. For instance, the result of the commands

```
t = 0 : 0.01 : 6;
x = square(t);
plot(t,x)
```

shows three square-wave cycles of length 2. (Try it.)

If $f(t)$ is one of the three period 2 engineering functions defined above, then the function $f(2t/p)$ will have period p. To illustrate this, try

```
p = 5;
t = 0 : p/200 : 3*p;
plot(t, sawtooth(2*t/p))
```

with various values of p. The resulting graph should show three sawtooth-wave cycles of length p. For instance, with $p = 5$ we get the period 5 sawtooth-wave graph shown at the top of the next page.

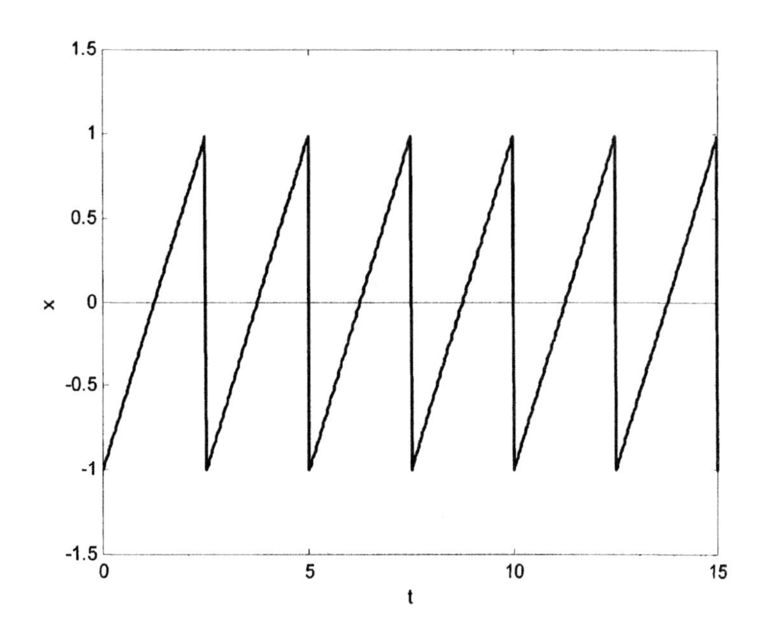

Now let's consider the mass-spring-dashpot equation in (1) with selected parameter values and an input forcing function having period p and amplitude F_0, for instance,

```
function   F = force(t)
p = 5;       % period
F0 = 400;    % amplitude
F = F0*sawtooth(2*t/p);
```

You can plot this input force function to verify that it has period 5:

```
t = 0 : 0.03 : 15;
plot(t, force(t))
```

Then our differential equation is defined (as a system of two first-order equations) by the function

```
function    xp = de(t,x)
m = 3;    c = 100;    k = 20;
xp = x;
y = x(2);
x = x(1);
xp(1) = y;
xp(2) = (-k*x - c*y + force(t))/m;
```

Next, let's suppose that the mass is initially at rest in its equilibrium position and solve numerically the resulting initial value problem.

```
[t,x] = ode45('de', 0:0.05:30, [0;0]);
response = x(:,1);
```

When we plot this solution, we see that after an initial transient dies out, the response function $x(t)$ settles down (as expected?) to a periodic oscillation with the same period as the input.

```
plot(t,response);
axis([0 30 -2.5 2.5])
hold on
plot([0 30],[0 0],'k')
```

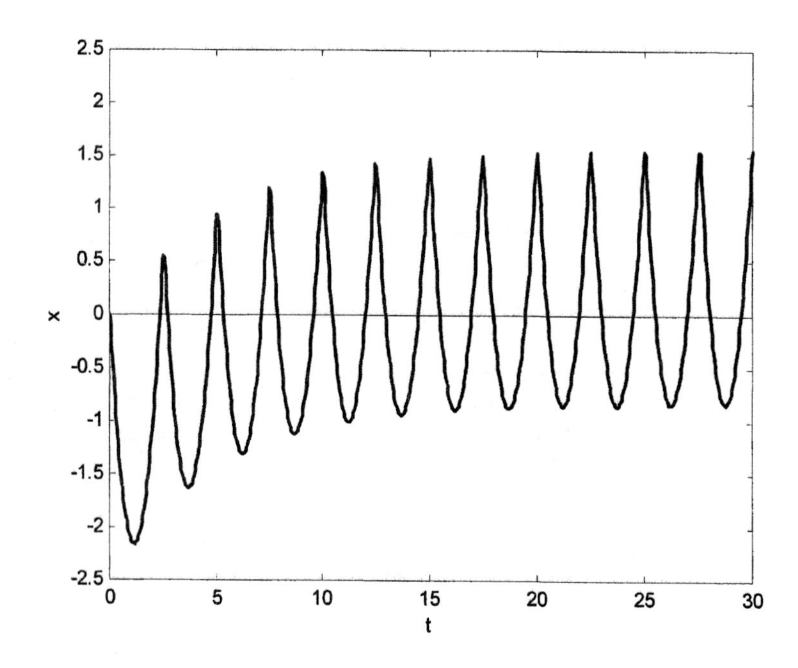

Chapter 8

Power Series Methods

Application 8.2
Automatic Computation of Series Coefficients

Repeated application of the recurrence relation to grind out successive coefficients is — especially in the case of a recurrence relation with three or more terms — a tedious aspect of the infinite series method. Here we illustrate the use of a computer algebra system for this task. In Example 7 of Section 8.2 in the text we saw that the coefficients in the series solution $y = \sum c_n x^n$ of the differential equation $y'' - x\,y' - x^2 y = 0$ are given in terms of the two arbitrary coefficients c_0 and c_1 by

$$c_2 = 0, \quad c_3 = \frac{c_1}{6}, \text{ and } c_{n+2} = \frac{n\,c_n + c_{n-2}}{(n+2)(n+1)} \quad \text{for } n \geq 2. \tag{1}$$

It looks like a routine matter to implement such a recurrence relation, but a twist results from the fact that a typical computer system array is indexed by the subscripts $1, 2, 3, \cdots$ rather than by the subscripts $0, 1, 2, \cdots$ that match the exponents in the successive terms of a power series that begins with a constant term. For this reason we rewrite our proposed power series solution in the form

$$y = \sum_{n=0}^{\infty} c_n x^n = \sum_{n=1}^{\infty} b_n x^{n-1} \tag{2}$$

where $b_n = c_{n-1}$ for each $n \geq 1$. Then the first two conditions in (1) say that $b_3 = 0,\ b_4 = b_2/6,$ and the recurrence relation (with n replaced with $n - 1$) yields the new recurrence relation

$$b_{n+2} = c_{n+1} = \frac{(n-1)c_{n-1} + c_{n-3}}{(n+1)(n)} = \frac{(n-1)b_n + b_{n-2}}{n(n+1)} \tag{3}$$

Now we're ready to go. The *Maple*, *Mathematica*, and MATLAB implementations of this recurrence relation illustrated in the paragraphs below yield the 11-term partial sum

$$y(x) = 1 + x + \frac{x^3}{6} + \frac{x^4}{12} + \frac{3x^5}{40} + \frac{x^6}{90} + \frac{13x^7}{1008} + \frac{3x^8}{1120} + \frac{119x^9}{51840} + \frac{41x^{10}}{113400} + \cdots. \tag{4}$$

You can apply this method to any of the examples and problems in Section 8.2 of the text.

Using *Maple*

Suppose we want to calculate the terms through the 10th degree — that is, 11 terms

```
k := 11:          # k terms
```

— in (2) with the initial conditions $b_1 = b_2 = 1$. We begin by setting up an array of k terms and specifying the given (initial) values of the first two terms.

```
b := array(1..k):
b[1] := 1:        # given  y(0) value
b[2] := 1:        # given y'(0) value
```

Also, because we know also that $b_3 = 0$ and $b_4 = b_2 / 6,$ we define the additional values

```
b[3] := 0:
b[4] := b[2]/6:
```

Using the recurrence relation in (3), subsequent coefficient values are now calculated by the loop

```
for n from 3 by 1 to k-2 do
    b[n+2] := ((n-1)*b[n]+b[n-2])/(n*(n+1));
    od;
```

which quickly yields the b-coefficient values corresponding to the solution given in (4). You might note that the even- and odd-degree terms there agree with those shown in Eqs. (18) and (19) of Example 7 in the text. You can substitute $b_1 = 1,\ b_2 = 0$ and $b_1 = 0,\ b_2 = 1$ separately (instead of $b_1 = b_2 = 1$) in the commands above to derive these two solutions separately.

Using *Mathematica*

Suppose we want to calculate the terms through the 10th degree — that is, 11 terms

```
k = 11;             (*  k terms  *)
```

— in (2) with the initial conditions $b_1 = b_2 = 1$. We begin by setting up an array of k terms and specifying the given (initial) values of the first two terms.

```
b = Table[0,{n,1,k}];
b[[1]] = 1;          (* given  y(0) value *)
b[[2]] = 1;          (* given y'(0) value *)
```

Also, because we know also that $b_3 = 0$ and $b_4 = b_2 / 6,$ we define the additional values

```
b[[3]] = 0;
b[[4]] = b[[2]]/6;
```

Using the recurrence relation in (3), subsequent coefficient values are now calculated by the loop

```
For[n = 3, n <= k-2,
    b[[n+2]] = ((n-1)b[[n]] + b[[n-2]])/(n*(n+1));
    n = n+1];
b
```

which quickly yields the *b*-coefficient values corresponding to the solution given in (4). You might note that the even- and odd-degree terms there agree with those shown in Eqs. (18) and (19) of Example 7 in the text. You can substitute $b_1 = 1,\ b_2 = 0$ and $b_1 = 0,\ b_2 = 1$ separately (instead of $b_1 = b_2 = 1$) in the commands above to derive these two solutions separately.

Using MATLAB

Suppose we want to calculate the terms through the 10th degree — that is, 11 terms

```
k = 11;        % k terms
```

— in (2) with the initial conditions $b_1 = b_2 = 1$. We begin by setting up an array of k terms and specifying the given (initial) values of the first two terms.

```
b = 0*(1:k);
b(1) = 1;    % given y(0) value
b(2) = 1;    % given y'(0) value
```

Also, because we know also that $b_3 = 0$ and $b_4 = b_2 / 6,$ we define the additional values

```
b(3) = 0;
b(4) = b(2)/6;
```

Using the recurrence relation in (3), subsequent coefficient values are now calculated by the loop

```
for n = 3:k-2
    b(n+2)  =  ((n-1)*b(n)+b(n-2))/(n*(n+1));
    end
```

When we display the resulting coefficient values

```
format rat, b
```

we see the same coefficients as displayed in the solution (4), *except that* the coefficient b_{10} of x^9 is shown as 73/31801 rather than the correct value 119/51840 shown in (4). It happens that

$$\frac{73}{31801} \approx 0.0022955253 \text{ while } \frac{119}{51840} \approx 0.0022955247$$

so the two rational fractions agree when rounded off to 9 decimal places. The explanation is that MATLAB (as opposed to *Maple* and *Mathematica*) works internally with decimal rather than exact arithmetic. But at the end of the computation, its **format rat** algorithm converts a correct 14-place approximation for b_{10} to an incorrect rational fraction that's "close but no cigar."

You can substitute $b_1 = 1,\ b_2 = 0$ and $b_1 = 0,\ b_2 = 1$ separately (instead of $b_1 = b_2 = 1$ in the commands above) to derive partial sums of the two linearly independent solutions displayed in Eqs. (18) and (19) of Example 7.

Further Investigation

Let's take things a step further by using the computer to illustrate Theorem 1 of Section 8.2. First, show that the coefficients in the series solution $y = \sum c_n x^n$ of the differential equation $(x^2+1)y'' + 6xy' + 4y = 0$ (taken from Problem 4 of the section) are given in terms of the two arbitrary coefficients c_0 and c_1 by $c_{n+2} = -\dfrac{(n+4)c_n}{(n+2)}$, $n \geq 0$. As above, we will write

$$y = \sum_{n=0}^{\infty} c_n x^n = \sum_{n=1}^{\infty} b_n x^{n-1},$$

where $b_n = c_{n-1}$ for each $n \geq 1$, leading to the new recurrence relation

$$b_{n+2} = c_{n+1} = -\frac{(n+3)c_{n-1}}{(n+1)} = -\frac{(n+3)b_n}{(n+1)}$$

First, we modify the code in the "Using *Mathematica*" section above to find the coefficients $b_1, b_2, \ldots, b_{11}$ for this differential equation with the same initial conditions $b_1 = b_2 = 1$. (*Maple* and MATLAB users can proceed similarly.) Then define the actual series approximation using

```
approx= Sum[b[[n]]*x^(n - 1), {n, 1, k}]
```

Our goal is to plot this approximation (in black) alongside the "true" solution (in red), as computed numerically by *Mathematica*'s **NDSolve** routine. Based on Theorem 1, over what interval of the x-axis would you expect the series approximation to agree with the "true" solution?

Use the following commands to test your answer:

```
temp = NDSolve[
            {(x^2 + 1)*y''[x] + 6x*y'[x] + 4y[x] -> 0,
            y[0] == -3, y'[0] -> -2}, y[x], {x,-2,2}];
true = y[x]/.temp

Plot[{approx, true}, {x,-2,2},
PlotStyle -> {GrayLevel[0],Hue[0]},PlotRange->{-2,2}];
```

Were your expectations confirmed? Now do it again, but this time using 20 terms in the series approximation, and then 100 terms; do things still come out as expected?

Application 8.3
Automating the Frobenius Series Method

Here we illustrate the use of a computer algebra system to apply the method of Frobenius. In the paragraphs that follow, we consider the differential equation

$$2x^2 y'' + 3x\, y' - (x^2 + 1)y = 0 \qquad (1)$$

of Example 4 in Section 8.3 of the text, where we found the two indicial roots $r_1 = \frac{1}{2}$ and $r_2 = -1$. We carry through the formal Frobenius method starting with the larger indicial root r_1, and you can then apply the same process to derive the second Frobenius series solution (found manually in the text) corresponding to r_2.

In the following examples, use this method to derive Frobenius series solutions that can be checked against the given known general solutions.

1. $x\, y'' - y' + 4x^3\, y = 0$, $\quad y(x) = A \cos x^2 + B \sin x^2$

2. $x\, y'' - 2\, y' + 9x^5\, y = 0$, $\quad y(x) = A \cos x^3 + B \sin x^3$

3 $4x\, y'' + 2\, y' + y = 0$, $\quad y(x) = A \cos \sqrt{x} + B \sin \sqrt{x}$

4. $x\, y'' + 2\, y' + x\, y = 0$, $\quad y(x) = \dfrac{1}{x}\left(A \cos x + B \sin x\right)$

5. $4x\, y'' + 6\, y' + y = 0$, $\quad y(x) = \dfrac{1}{\sqrt{x}}\left(A \cos \sqrt{x} + B \sin \sqrt{x}\right)$

6. $x\, y'' + x\, y' + (4x^4 - 1)\, y = 0$, $y(x) = \dfrac{1}{x}\left(A \cos x^2 + B \sin x^2\right)$

7. $x\, y'' + 3\, y' + 4x^3\, y = 0$, $\quad y(x) = \dfrac{1}{x^2}\left(A \cos x^2 + B \sin x^2\right)$

8. $x^2\, y'' + x^2\, y' - 2\, y = 0$, $\quad y(x) = \dfrac{1}{x}\left[A\,(2 - x) + B\,(2 + x)\, e^{-x}\right]$

The next three problems involve the arctangent series

$$\tan^{-1} x = x - \frac{x^3}{3} + \frac{x^5}{5} - \frac{x^7}{7} + \cdots .$$

9. $(x+x^3)\, y'' + (2+4x^2)\, y' - 2x\, y = 0, \qquad y(x) = \frac{1}{x}\left(A + B \tan^{-1} x\right)$

10. $(2x+2x^2)\, y'' + (3+5x)\, y' + y = 0, \quad y(x) = \frac{1}{\sqrt{x}}\left(A + B \tan^{-1} \sqrt{x}\right)$

11. $(x+x^5)\, y'' + (3+7x^4)\, y' + 8x^3\, y = 0, \qquad y(x) = \frac{1}{x^2}\left(A + B \tan^{-1} x^2\right)$

Using *Maple*

Beginning with the indicial root

```
r := 1/2:
```

we first write the initial k+1 terms of a proposed Frobenius series solution:

```
k := 6:
a := array(0..k):
y := x^(1/2)*sum( a[n]*x^(n), n = 0..k);
```

$$y := \sqrt{x}\left(a_0 + a_1 x + a_2 x^2 + a_3 x^3 + a_4 x^4 + a_5 x^5 + a_6 x^6\right)$$

Then we substitute this series (partial sum) into the left-hand side of Equation (1),

```
de := 2*x^2*diff(y,x$2) + 3*x*diff(y,x) - (x^2+1)*y:
eq1 := simplify(de);
```

$$deq1 := -x^{\left(\frac{3}{2}\right)}\left(-5a_1 - 14a_2x - 27a_3x^2 - 44a_4x^3 - 65a_5x^4 - 90a_6x^5 + a_0x + x^6a_5 + x^7a_6 + x^2a_1 + x^3a_2 + x^4a_3 + x^5a_4\right)$$

Noting the $x^{3/2}$ factor, we multiply by $x^{-3/2}$ and then collect coefficients of like powers of x.

```
eq2 := collect( x^(-3/2)*eq1, x);
```

$$deq2 := -x^7a_6 - x^6a_5 + (-a_4 + 90a_6)x^5 + (-a_3 + 65a_5)x^4 + (-a_2 + 44a_4)x^3 + (-a_1 + 27a_3)x^2 + (14a_2 - a_0)x + 5a_1$$

Next we set up the equations the successive coefficients must satisfy. We do this by defining an array and then filling the elements of this array by equating (in turn) each of the series coefficients to zero.

```
eqs := array(0..5):
for n from 0 to 5 do
     eqs[n] := coeff(eq2,x,n) = 0:
     od:
coeffEqs := convert(eqs, set);
```

$$coeffEqs := \{\, 5\,a_1 = 0, -a_2 + 44\,a_4 = 0, -a_3 + 65\,a_5 = 0,\\ 90\,a_6 - a_4 = 0, 14\,a_2 - a_0 = 0, -a_1 + 27\,a_3 = 0 \,\}$$

We have here a collection of six linear equations relating the seven coefficients a_0 through a_6. Hence we should be able to solve for the successive coefficients

```
succCoeffs := convert([seq(a[n], n=1..6)], set);
```

$$succCoeffs := \{\, a_1, a_2, a_3, a_4, a_5, a_6 \,\}$$

in terms of a_0:

```
ourCoeffs := solve(coeffEqs, succCoeffs);
```

$$ourCoeffs := \left\{ a_1 = 0, a_3 = 0, a_5 = 0, a_2 = \frac{1}{14} a_0, a_4 = \frac{1}{616} a_0, a_6 = \frac{1}{55440} a_0 \right\}$$

Finally we substitute all these coefficients back into the original series.

```
partSoln := subs(ourCoeffs, y);
```

$$partSoln := \sqrt{x}\left(a_0 + \frac{1}{14} a_0\, x^2 + \frac{1}{616} a_0\, x^4 + \frac{1}{55440} a_0\, x^6 \right)$$

Note that (after factoring out a_0) this result agrees with the first particular solution

$$y_1(x) = a_0 x^{1/2}\left(1 + \frac{x^2}{14} + \frac{x^4}{616} + \frac{x^6}{55440} + \cdots \right)$$

found in the text.

Using *Mathematica*

Beginning with the indicial root

```
r = 1/2;
```

we first write the initial $k+1$ terms of a proposed Frobenius series solution:

```
k = 6;

y = x^r (Sum[ a[n] x^n, {n,0,k} ] + O[x]^(k+1))
```

$$a(0)\sqrt{x} + a(1)x^{3/2} + a(2)x^{5/2} + a(3)x^{7/2} + a(4)x^{9/2} + a(5)x^{11/2} + a(6)x^{13/2} + O(x^{15/2})$$

Then we substitute this series into the differential equation in (1),

```
deq =
2x^2 D[y, x,x] + 3x D[y, x] - (x^2 + 1)y == 0
```

$$5a(1)x^{3/2} + (14a(2) - a(0))x^{5/2} + (27a(3) - a(1))x^{7/2} + (44a(4) - a(2))x^{9/2} + (65a(5) - a(3))x^{11/2} + (90a(6) - a(4))x^{13/2} + O(x^{15/2}) == 0$$

Mathematica has automatically collected like powers for us, and we can use the **`LogicalExpand`** command to extract the equations that the successive coefficients satisfy.

```
coeffEqns   =   LogicalExpand[ deq ]
```

$$5a(1) == 0 \wedge 14a(2) - a(0) == 0 \wedge 27a(3) - a(1) == 0 \wedge 44a(4) - a(2) == 0 \wedge 65a(5) - a(3) == 0 \wedge 90a(6) - a(4) == 0$$

We have here a collection of six linear equations relating the seven coefficients $a_0 = a(0)$ through $a_6 = a(6)$. Hence we should be able to solve for the successive coefficients

```
succCoeffs   =   Table[ a[n], {n, 1, 6} ]
```

$$\{a(1), a(2), a(3), a(4), a(5), a(6)\}$$

in terms of $a(0)$:

```
ourCoeffs   =   Solve[coeffEqns, succCoeffs]
```

$$\left\{\left\{a(1) \to 0, a(2) \to \frac{a(0)}{14}, a(3) \to 0, a(4) \to \frac{a(0)}{616}, a(5) \to 0, a(6) \to \frac{a(0)}{55440}\right\}\right\}$$

Finally we substitute all these coefficients back into the original series.

```
partSoln = y /. ourCoeffs // First
```

$$a(0)\sqrt{x}+\frac{1}{14}a(0)x^{5/2}+\frac{1}{616}a(0)x^{9/2}+\frac{a(0)x^{13/2}}{55440}+O(x^{15/2})$$

Note that (after factoring out $a_0x^{1/2}$) this result agrees with the first particular solution

$$y_1(x) = a_0x^{1/2}\left(1+\frac{x^2}{14}+\frac{x^4}{616}+\frac{x^6}{55440}+\cdots\right).$$

found in the text.

Using MATLAB

Beginning with the indicial root $r_1 = \frac{1}{2}$, we first write the initial seven terms of a proposed Frobenius series solution:

```
syms x y a0 a1 a2 a3 a4 a5 a6
y =
x^(1/2)*(a0+a1*x+a2*x^2+a3*x^3+a4*x^4+a5*x^5+a6*x^6);
pretty(y)
```

```
 1/2                       2       3       4       5       6
x    (a0 + a1 x + a2 x  + a3 x  + a4 x  + a5 x  + a6 x )
```

Then we substitute this series (partial sum) into the left-hand side of Equation (1):

```
de = 2*x^2*diff(y,2) + 3*x*diff(y) - (x^2 + 1)*y;
simple(de)
ans =
-x^(3/2)*(x^6*a5+x^7*a6+x*a0+x^4*a3+x^3*a2+x^2*a1
            +x^5*a4-65*a5*x^4-44*a4*x^3-27*a3*x^2
            -14*a2*x-5*a1-90*a6*x^5)
```

Noting the $x^{3/2}$ factor, we multiply by $x^{-3/2}$ and then collect coefficients of like powers of x.

```
de = collect(simplify(x^(-3/2)*de))
de =
-x^7*a6-x^6*a5+ (90*a6-a4)*x^5 + (-a3+65*a5)*x^4
 + (-a2+44*a4)*x^3 + (-a1+27*a3)*x^2
 + (14*a2-a0)*x + 5*a1
```

Next we set up the equations the original coefficients in our Frobenius series must satisfy. By cutting and pasting it is a simple matter to pick out the successive coefficients in **de** and equate each of them to zero.

```
eq1 = 5*a1;
eq2 = 14*a2-a0;
eq3 = -a1+27*a3;
eq4 = -a2+44*a4;
eq5 = -a3+65*a5;
eq6 = 90*a6-a4;
```

We have here a collection of six linear equations relating the seven coefficients **a0** through **a6**. Hence we should be able to solve for the successive coefficients **a1** through **a6** in terms of **a0**:

```
soln =
solve(eq1,eq2,eq3,eq4,eq5,eq6, a1,a2,a3,a4,a5,a6);
coeffs =
[soln.a1,soln.a2,soln.a3,soln.a4,soln.a5,soln.a6]
coeffs =
[ 0,   1/14*a0,   0,   1/616*a0,   0,   1/55440*a0]
```

Finally we substitute all these coefficients back into the original series **y**:

```
y = subs(y, [a1,a2,a3,a4,a5,a6], coeffs);
pretty(y)
 1/2                      2                4                  6
x     (a0 + 1/14 a0 x  + 1/616 a0 x  + 1/55440 a0 x )
```

Note that (after factoring out a_0) this result agrees with the first particular solution

$$y_1(x) = a_0 x^{1/2}\left(1+\frac{x^2}{14}+\frac{x^4}{616}+\frac{x^6}{55440}+\cdots\right).$$

found in the text.

Application 8.4
The Exceptional Case by Reduction of Order

The reduction of order formula in Eq. (28) of Section 8.4 in the text is readily implemented using a computer algebra system. Recall that, given a known solution $y_1(x)$ of the homogeneous linear second-order equation

$$y'' + P(x)\,y' + Q(x)\,y = 0 \tag{1}$$

on an open interval where P and Q are continuous, the **reduction of order formula** gives a second independent solution $y_2(x)$ defined by

$$y_2(x) = y_1(x)\int \frac{\exp(-\int P(x)\,dx)}{y_1(x)^2}\,dx. \tag{2}$$

In the paragraphs below we illustrate the use of (2) with *Maple*, *Mathematica*, and MATLAB to derive a second solution of Bessel's equation of order zero, beginning with the known power series solution

$$J_0(x) = \sum_{n=0}^{\infty}(-1)^n \frac{x^{2n}}{2^{2n}\,(n!)^2}. \tag{3}$$

After verifying (with your computer algebra system) the computations we present, you can begin with the power series for $J_1(x)$ in Eq. (49) of Section 8.4 and derive similarly the second solution in (50) of Bessel's equation of order 1. Problems 9 through 14 in Section 8.4 can also be partially automated in this way.

Using *Maple*

We begin by entering the initial terms of the known series solution (3) in the form

```
y1 := sum((-x^2/4)^n / (n!)^2, n=0..5);
```

$$y1 := 1 - \frac{1}{4}x^2 + \frac{1}{64}x^4 - \frac{1}{2304}x^6 + \frac{1}{147456}x^8 - \frac{1}{14745600}x^{10}$$

Then we need only start with

```
P := 1/x:
```

and "build" the integral that appears in Eq. (2):

```
part1 := exp(-int(P,x));
```

$$part1 := \frac{1}{x}$$

```
part2 := series(1/y1^2, x, 11):
part2 := convert(part2, polynom);
```

$$part2 := 1 + \frac{1}{2}x^2 + \frac{5}{32}x^4 + \frac{23}{5764}x^6 + \frac{677}{73728}x^8 + \frac{7313}{3686400}x^{10}$$

Now we're ready to evaluate the integral in the reduction of order formula:

```
integral := int(part1*part2, x);
```

$$integral := \frac{7313}{36864000}x^{10} + \frac{677}{589824}x^8 + \frac{23}{3456}x^6 + \frac{5}{128}x^4 + \frac{1}{4}x^2 + \ln(x)$$

Finally, we first multiply **y1** times the series part of this integral,

```
y := series(y1*(integral - ln(x)), x, 11);
```

$$y := \frac{1}{4}x^2 - \frac{3}{128}x^4 + \frac{11}{13824}x^6 - \frac{25}{1769472}x^8 + \frac{137}{884736000}x^{10} + O(x^{12})$$

and then add on **y1 = J0** times the logarithmic term in the integral:

```
y2 := J0*ln(x) + y;
```

$$y2 := J0\ln(x) + \left(\frac{1}{4}x^2 - \frac{3}{128}x^4 + \frac{11}{13824}x^6 - \frac{25}{1769472}x^8 + \frac{137}{884736000}x^{10} + O(x^{12})\right)$$

Thus we obtain the second solution of Bessel's equation of order 0 (as we see it in Eq. (45) in Section 8.4 of the text).

Using *Mathematica*

We begin by entering the initial terms of the known series solution (3) in the form

```
y1  =  Sum[(-x^2/4)^n /(n!)^2, {n,0,5}] + O[x]^12
```

$$y1 := 1 - \frac{x^2}{4} + \frac{x^4}{64} - \frac{x^6}{2304} + \frac{x^8}{147456} - \frac{x^{10}}{14745600} + O(x^{12})$$

Then we need only start with

```
P  =  1/x;
```

and evaluate the integral that appears in the reduction of order formula (2):

```
integral = Integrate[Exp[-Integrate[P,x]]/y1^2, x]
```

$$\log(x)+\frac{x^2}{4}+\frac{5x^4}{128}+\frac{23x^6}{3456}+\frac{677x^8}{589824}+\frac{7313x^{10}}{36864000}+O(x^{12})$$

Finally, we first multiply **y1** times the series part of this integral,

```
y = y1 (integral - Log[x])
```

$$\frac{x^2}{4}-\frac{3x^4}{128}+\frac{11x^6}{13824}-\frac{25x^8}{1769472}+\frac{137x^{10}}{884736000}+O(x^{12})$$

and then add on **y1 = J0** times the logarithmic term in the integral:

```
y2  =  J0 Log[x] + y
```

$$\text{J0}\log(x)+\frac{x^2}{4}-\frac{3x^4}{128}+\frac{11x^6}{13824}-\frac{25x^8}{1769472}+\frac{137x^{10}}{884736000}+O(x^{12})$$

Thus we obtain the second solution of Bessel's equation of order 0 (as we see it in Eq. (45) in Section 8.4 of the text).

Using MATLAB

We begin by entering the initial terms of the known series solution (3) in the form

```
syms x
y1 = 1 - x^2/4 + x^4/64 - x^6/2304 + x^8/147456 ...
       - x^10/14745600;
pretty(y1)
```

```
          2           4             6               8                  10
1 - 1/4 x  + 1/64 x  - 1/2304 x  + 1/147456 x  - 1/14745600 x
```

Then we need only start with

```
P = 1/x;
```

and "build" the integral that appears in the reduction of order formula (2). Using the **taylor** function to compute series expansions, we calculate

```
part1 = simplify(exp(-int(P)))
part1 =
     1/x
```

```
part2 = taylor(1/(y1^2),12);
pretty(part2)

              2            4    23    6     677    8      7313      10
1 + 1/2 x   + 5/32 x   + --- x   + ----- x   + ------- x
                          576       73728      3686400
```

Now we're ready to evaluate the integral in the reduction of order formula:

```
integral = int(part1*part2);
pretty(integral)

  7313     10     677      8     23    6              4          2
-------- x   + ------ x   + ---- x   + 5/128 x   + 1/4 x   + log(x)
36864000        589824      3456
```

Finally, we first multiply **y1** times the series part of this integral,

```
y = taylor(y1*(integral-log(x)), 12);
pretty(y)

    2            4     11     6       25      8         137       10
1/4 x   - 3/128 x   + ----- x   - ------- x   + --------- x
                      13824       1769472       884736000
```

and then add on **y1 = J0** times the logarithmic term in the integral:

```
y2 = J0*log(x)+y;
pretty(y2)

                 2          4    11    6       25     8       137      10
J0 log(x)+1/4 x   -3/128 x   + -----x   - -------x + ---------x
                               13824     1769472    884736000
```

Thus we obtain the second solution of Bessel's equation of order 0 (as we see it in Eq. (45) in Section 8.4 of the text).

Application 8.6
Riccati Equations and Modified Bessel Functions

A Riccati equation is a first-order differential equation of the form

$$y' = A(x)\,y^2 + B(x)\,y + C(x)$$

(with a single nonlinear $A\,y^2$ term). Many Riccati equations like the ones listed below can be solved explicitly in terms of Bessel functions.

$$y' = x^2 + y^2 \tag{1}$$

$$y' = x^2 - y^2 \tag{2}$$

$$y' = y^2 - x^2 \tag{3}$$

$$y' = x + y^2 \tag{4}$$

$$y' = x - y^2 \tag{5}$$

$$y' = y^2 - x \tag{6}$$

For instance, Problem 15 in Section 8.6 of the text says that the general solution of (1) is given by

$$y(x) = x \cdot \frac{J_{3/4}\left(\tfrac{1}{2}x^2\right) - c\,J_{-3/4}\left(\tfrac{1}{2}x^2\right)}{c\,J_{1/4}\left(\tfrac{1}{2}x^2\right) + J_{-1/4}\left(\tfrac{1}{2}x^2\right)}. \tag{7}$$

See whether the symbolic DE solver command in your computer algebra system, such as

```
dsolve(diff(y(x),x) = x^2 + y(x)^2, y(x))
```
(*Maple*)

or

```
DSolve[y'[x] == x^2 + y[x]^2, y[x], x]
```
(*Mathematica*)

or

```
dsolve('Dy = x^2 + y^2')
```
(MATLAB)

agrees with (7).

If Bessel functions other than those appearing in (7) are involved, you may need to apply Identities (26) and (27) in Section 8.5 of the text to transform the computer's "answer" to (7). Then see whether your system can take the limit as $x \to 0$ in (7) to show that the arbitrary constant c is given in terms of the initial value $y(0)$ by

$$c = -\frac{y(0)\,\Gamma\left(\frac{1}{4}\right)}{2\,\Gamma\left(\frac{3}{4}\right)} \tag{8}$$

Now you should be able to use built-in Bessel functions to plot typical solution curves like those illustrated below.

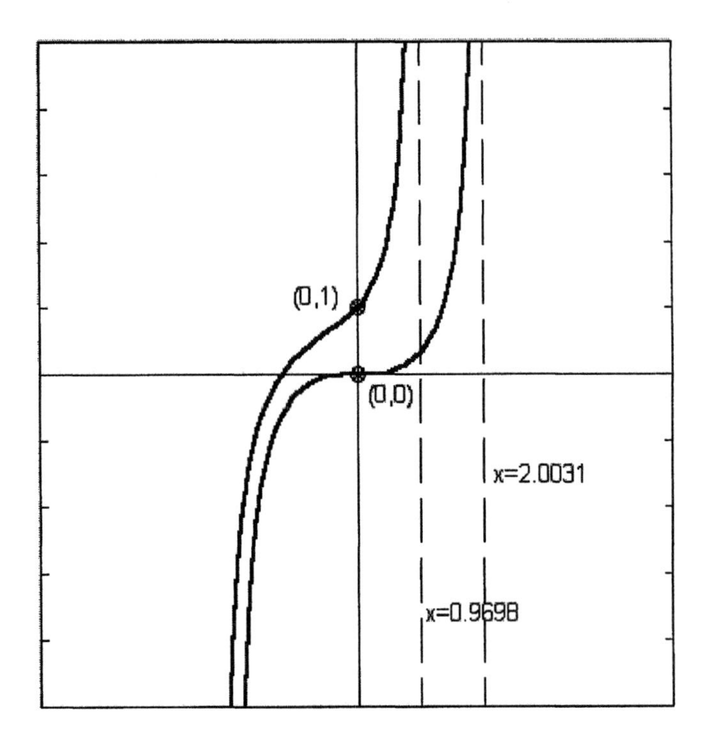

This figure shows the trajectories of (1) satisfying the initial conditions $y(0)=0$ and $y(0)=1$, together with apparent vertical asymptotes. Can you use (7) and (8) to verify that these asymptotes are given approximately by $x = 2.00315$ and $x = 0.96981$, respectively? (*Suggestion*: You are looking for zeros of the denominator in (7).)

Next, investigate similarly one of the other equations in (2)–(6). Each has a general solution of the same general form as in (7) — a quotient of linear combinations of Bessel functions. In addition to $J_p(x)$ and $Y_p(x)$, these solutions may involve the *modified Bessel functions*

$$I_p(x) = i^{-p} J_p(ix)$$

and

$$K_p(x) = \frac{\pi}{2} i^{-p} \left[J_p(ix) + Y_p(ix) \right]$$

that satisfy the *modified Bessel equation*

$$x^2 y'' + x\,y' - (x^2 + p^2)y = 0$$

of order p. For instance, the general solution of Equation (5) is given for $x > 0$ by

$$y(x) = \sqrt{x} \cdot \frac{I_{2/3}\left(\frac{2}{3}x^{3/2}\right) - c\,I_{-2/3}\left(\frac{2}{3}x^{3/2}\right)}{I_{-1/3}\left(\frac{2}{3}x^{3/2}\right) - c\,I_{1/3}\left(\frac{2}{3}x^{3/2}\right)} \tag{9}$$

where

$$c = -\frac{y(0)\,\Gamma\left(\frac{1}{3}\right)}{\sqrt[3]{3}\,\Gamma\left(\frac{2}{3}\right)}. \tag{10}$$

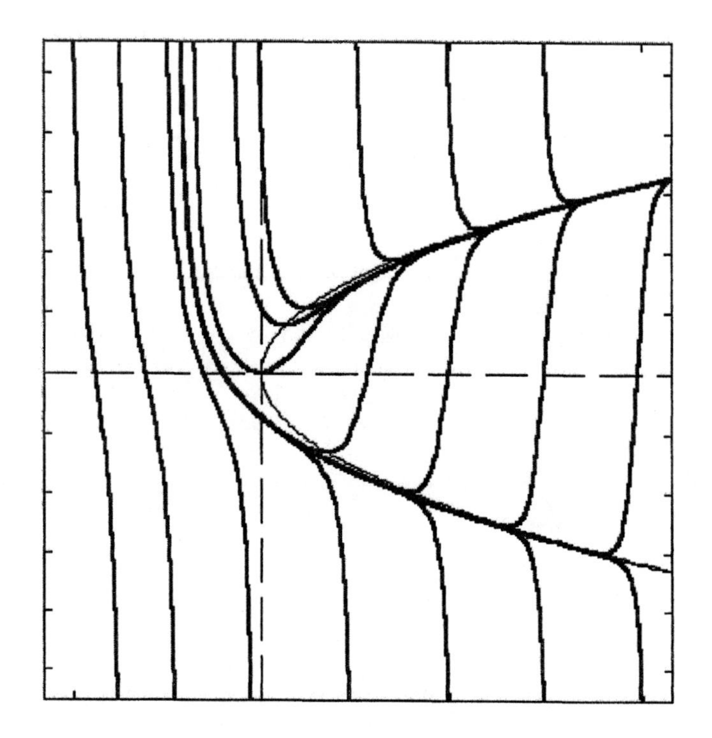

The figure above shows some typical solution curves for Equation (5), together with the parabola $y^2 = x$ that appears to bear an interesting relation to these curves — we see a funnel near $y = +\sqrt{x}$ and a spout near $y = -\sqrt{x}$.

The Bessel functions with imaginary argument that appear in the definitions of $I_p(x)$ and $K_p(x)$ may look exotic, but the power series of the modified function

$I_n(x)$ is simply that of the unmodified function $J_n(x)$, except without the alternating minus signs. For instance,

$$I_0(x) = 1 + \frac{x^2}{4} + \frac{x^4}{64} + \frac{x^6}{2304} + \cdots$$

and

$$I_1(x) = \frac{x}{2} + \frac{x^3}{16} + \frac{x^5}{384} + \frac{x^7}{18432} + \cdots.$$

Check these power series expansions using your computer algebra system — look at **BesselI** in either *Maple* or *Mathematica* — and compare them with Eqs. (17)–(18) in Section 8.5 of the text.

In the paragraphs that follow we illustrate the use of *Maple*, *Mathematica*, and MATLAB to investigate solutions of the Riccati equation $y' = x^2 + y^2$ in Eq. (1).

Using *Maple*

The general solution of the Riccati equation

```
de :=    diff(y(x),x) = x^2 + y(x)^2:
```

is given by

```
soln := dsolve(de, y(x)):
y := subs(_C1=c, rhs(soln));
```

$$y0 := -\frac{x\left(c\ \mathrm{BesselJ}\left(\frac{-3}{4}, \frac{1}{2}x^2\right) + \mathrm{BesselY}\left(\frac{-3}{4}, \frac{1}{2}x^2\right)\right)}{c\ \mathrm{BesselJ}\left(\frac{1}{4}, \frac{1}{2}x^2\right) + \mathrm{BesselY}\left(\frac{1}{4}, \frac{1}{2}x^2\right)}$$

Note that this form of the solution differs from (7) in that it involves the Bessel functions $Y_{-3/4}$ and $Y_{1/4}$ of the second kind rather than the Bessel functions $J_{-3/4}$ and $J_{-1/4}$ of the first kind.

In order to impose an initial condition, we must therefore evaluate the limit as $x \to 0$ instead of using (8). The left-hand and right-hand limits of $y(x)$ at $x = 0$ are given by

```
limit(y, x=0, right);
```

$$\frac{\Gamma\left(\frac{3}{4}\right)^2 (c+1)}{\pi}$$

```
limit(y, x=0, left);
```

$$-\frac{\Gamma\left(\frac{3}{4}\right)^2 (c+1)}{\pi}$$

We therefore see that the particular solution satisfying the initial condition $y(0) = 0$ is obtained with the arbitrary constant value $c = -1$:

```
y0 := subs(c=-1, y);
```

$$y0 := -\frac{x\left(-\text{BesselJ}\left(\frac{-3}{4}, \frac{1}{2}x^2\right) + \text{BesselY}\left(\frac{-3}{4}, \frac{1}{2}x^2\right)\right)}{-\text{BesselJ}\left(\frac{1}{4}, \frac{1}{2}x^2\right) + \text{BesselY}\left(\frac{1}{4}, \frac{1}{2}x^2\right)}$$

When we graph this particular solution $y_0(x)$,

```
plot(y0, x=-2..2, -5..5);
```

we see an apparent vertical asymptote near $x = 2$. This vertical asymptote corresponds to a zero of the denominator of $y_0(x)$, which we proceed to locate:

```
denom(y0);
```

$$\text{BesselJ}\left(\frac{1}{4}, \frac{1}{2}x^2\right) - \text{BesselY}\left(\frac{1}{4}, \frac{1}{2}x^2\right)$$

```
fsolve(denom(y0)=0, x, 1.9..2.1);
```

$$2.003147359$$

Thus this vertical asymptote is given by $x \approx 2.0031$.

When you investigate similarly the particular solution $y_1(x)$ satisfying the initial condition $y(0) = 1$, note (from the left-hand and right-hand limits above) that different values of the arbitrary constant c must be chosen for $x < 0$ and for $x > 0$.

Using *Mathematica*

The general solution of the Riccati equation

```
de =   y'[x] == x^2 + y[x]^2;
```

is given by

```
soln = DSolve[de, y[x], x];

soln =
y[x] /. (soln /. {(x^4)^(1/2) -> x^2}) // First
```

$$\frac{J_{-\frac{3}{4}}\left(\frac{x^2}{2}\right)x^2 - J_{\frac{5}{4}}\left(\frac{x^2}{2}\right)x^2 + J_{-\frac{5}{4}}\left(\frac{x^2}{2}\right)c_1x^2 - J_{\frac{3}{4}}\left(\frac{x^2}{2}\right)c_1x^2 + J_{\frac{1}{4}}\left(\frac{x^2}{2}\right) + J_{\frac{1}{4}}\left(\frac{x^2}{2}\right)c_1}{-2\,x\,J_{\frac{1}{4}}\left(\frac{x^2}{2}\right) - 2\,x\,J_{-\frac{1}{4}}\left(\frac{x^2}{2}\right)c_1}$$

To "get rid of" the Bessel functions of order $\pm 5/4$ we need to apply the reduction rules given in Eqs. (26) and (27) of Section 8.5 in the text:

```
rule =
BesselJ[p+s,x] -> (2p/x)BesselJ[p,x] - BesselJ[p-s,x];

rule1 = rule /. {p -> 1/4, s -> 1, x -> x^2/2}
```

$$J_{\frac{5}{4}}\left(\frac{x^2}{2}\right) \to \frac{J_{\frac{1}{4}}\left(\frac{x^2}{2}\right)}{x^2} - J_{-\frac{3}{4}}\left(\frac{x^2}{2}\right)$$

```
rule2 = rule /. {p -> -(1/4), s -> -1, x -> x^2/2}
```

$$J_{-\frac{5}{4}}\left(\frac{x^2}{2}\right) \to \frac{J_{-\frac{1}{4}}\left(\frac{x^2}{2}\right)}{x^2} - J_{\frac{3}{4}}\left(\frac{x^2}{2}\right)$$

When we make these substitutions and "adjust" the arbitrary constant notation, we get the following simple form

```
soln =
soln /. {rule1, rule2, C[1] -> 1/c}] // Simplify
```

$$\frac{x\left(J_{\frac{3}{4}}\left(\frac{x^2}{2}\right) - c\,J_{-\frac{3}{4}}\left(\frac{x^2}{2}\right)\right)}{c\,J_{\frac{1}{4}}\left(\frac{x^2}{2}\right) + J_{-\frac{1}{4}}\left(\frac{x^2}{2}\right)}$$

that agrees with the general solution given in Eq. (7) above.

In order to impose an initial condition, we evaluate the limit as $x \to 0$ explicitly (instead of merely using (8)). The left-hand and right-hand limits of $y(x)$ at $x = 0$ are given by

```
y0 =
Limit[soln, x -> 0, Direction->1]   (* with  x < 0 *)
```

$$\frac{2c\Gamma\left(\frac{3}{4}\right)}{\Gamma\left(\frac{1}{4}\right)}$$

```
y0 =
Limit[soln, x -> 0, Direction->-1]   (* with  x > 0 *)
```

$$-\frac{2c\Gamma\left(\frac{3}{4}\right)}{\Gamma\left(\frac{1}{4}\right)}$$

Thus we must use different values of c for $x < 0$ and for $x > 0$. Using the second value of y_0 above, the value of c for $x > 0$ is given by

```
c = c /. Solve[ y0 == 1, c] // First
```

$$-\frac{\Gamma\left(\frac{1}{4}\right)}{2\Gamma\left(\frac{3}{4}\right)}$$

Finally, then, the particular solution (for $x > 0$) of Eq. (1) such that $y(0) = 1$ is

```
y1 = soln // Simplify
```

$$\frac{x\left(\Gamma\left(\frac{1}{4}\right)J_{-\frac{3}{4}}\left(\frac{x^2}{2}\right)+2\,\Gamma\left(\frac{3}{4}\right)J_{\frac{3}{4}}\left(\frac{x^2}{2}\right)\right)}{\Gamma\left(\frac{1}{4}\right)J_{\frac{1}{4}}\left(\frac{x^2}{2}\right)-2\,\Gamma\left(\frac{3}{4}\right)J_{-\frac{1}{4}}\left(\frac{x^2}{2}\right)}$$

When we graph this particular solution $y_1(x)$,

```
Plot[ y1, {x,0,1}, PlotRange -> {0,10}];
```

we see an apparent vertical asymptote near $x = 1$. This vertical asymptote corresponds to a zero of the denominator of $y_1(x)$, which we proceed to locate:

```
Denominator[y1]
```

$$\Gamma\left(\frac{1}{4}\right)J_{\frac{1}{4}}\left(\frac{x^2}{2}\right)-2\,\Gamma\left(\frac{3}{4}\right)J_{-\frac{1}{4}}\left(\frac{x^2}{2}\right)$$

```
FindRoot[Denominator[y1]==0, {x,1}]
```

$\{x \to 0.969811\}$

Thus this vertical asymptote is given by $x \approx 0.9698$.

Using MATLAB

The solution $y_0(x)$ of the initial value problem

$$y' = x^2 + y^2, \qquad y(0) = 0$$

is given by

```
y0 = dsolve('Dy = x^2 + y^2','y(0)=0','x')
pretty(y0)
                                 2                                2
    x (-besselj(-3/4, 1/2 x ) + bessely(-3/4, 1/2 x ))
 - ---------------------------------------------------------
                               2                               2
       -besselj(1/4, 1/2 x ) + bessely(1/4, 1/2 x )
```

When we plot this particular solution,

```
ezplot(y0, [0 3])
```

we see an apparent vertical asymptote located near $x = 2$. This vertical asymptote corresponds to a zero of the denominator of $y_0(x)$, which we proceed to locate:

```
[num, den] = numden(y0);
den
den =
besselj(1/4,1/2*x^2)-bessely(1/4,1/2*x^2)

fzero('besselj(1/4,1/2*x^2)-bessely(1/4,1/2*x^2)', 2)
ans =
    2.0031
```

Thus this vertical asymptote is given by $x \approx 2.0031$.

The solution $y_1(x)$ of the initial value problem

$$y' = x^2 + y^2, \qquad y(0) = 1$$

is given by

```
y1 = dsolve('Dy = x^2 + y^2','y(0)=1','x');
```

A fairly simple form for this particular solution $y_1(x)$ is given by

```
y1 = simplify(y1);
pretty(y1)

-x (-besselj(-3/4, 1/2 x ) gamma(3/4)  + bessely(-3/4, 1/2 x ) gamma(3/4)

                          2     /                          2            2
    + pi besselj(-3/4, 1/2 x ))  /  (-besselj(1/4, 1/2 x ) gamma(3/4)
                                /

                      2           2                         2
    + bessely(1/4, 1/2 x ) gamma(3/4)  + pi besselj(1/4, 1/2 x ))
```

When we graph this solution,

```
ezplot(y1, [0 1])
```

we see an apparent vertical asymptote located near $x = 1$.

The observed vertical asymptote to the graph $y = y_1(x)$ corresponds to a zero of the denominator of $y_1(x)$, which we proceed to locate:

```
[num, den] = numden(y1);
den
den =
-besselj(1/4,1/2*x^2)*gamma(3/4)^2+
bessely(1/4,1/2*x^2)*gamma(3/4)^2+
pi*besselj(1/4,1/2*x^2)

fzero('-besselj(1/4,1/2*x^2)*gamma(3/4)^2+
        bessely(1/4,1/2*x^2)*gamma(3/4)^2+
        pi*besselj(1/4,1/2*x^2)', 1)
ans =
    0.9698
```

Thus this vertical asymptote is given by $x \approx 0.9698$.

Chapter 9

Fourier Series Methods

Application 9.2
Computer Algebra Calculation of Fourier Coefficients

A computer algebra system can greatly ease the burden of calculation of the Fourier coefficients of a given function $f(t)$. In the case of a function defined "piecewise," we must take care to "split" the integral according to the different intervals of definition of the function. In the paragraphs that follow we illustrate the use of *Maple*, *Mathematica*, and MATLAB in deriving the Fourier series

$$f(t) = \frac{4}{\pi}\sum_{n\,\text{odd}}^{\infty}\frac{\sin nt}{n} \tag{1}$$

of the period 2π square wave function defined on $(-\pi, \pi)$ by

$$f(t) = \begin{cases} -1 & \text{if } -\pi < t < 0, \\ +1 & \text{if } 0 < t < \pi. \end{cases} \tag{2}$$

In this case the function is defined by different formulas on two different intervals, so each Fourier coefficient integral from $-\pi$ to π must be calculated as the sum of two integrals:

$$\begin{aligned} a_n &= \tfrac{1}{\pi}\int_{-\pi}^{0}(-1)\cos nt\,dt + \tfrac{1}{\pi}\int_{0}^{\pi}(+1)\cos nt\,dt, \\ b_n &= \tfrac{1}{\pi}\int_{-\pi}^{0}(-1)\sin nt\,dt + \tfrac{1}{\pi}\int_{0}^{\pi}(+1)\sin nt\,dt. \end{aligned} \tag{3}$$

To practice the symbolic derivation of Fourier series in this manner, you can begin by verifying the Fourier series calculated manually in Examples 1 and 2 of Section 9.2 in the text. Then Problems 1 through 21 there are fair game. Finally, the period 2π triangular wave and trapezoidal wave functions illustrated in the figures at the top of the next page have especially interesting Fourier series that we invite you to discover for yourself.

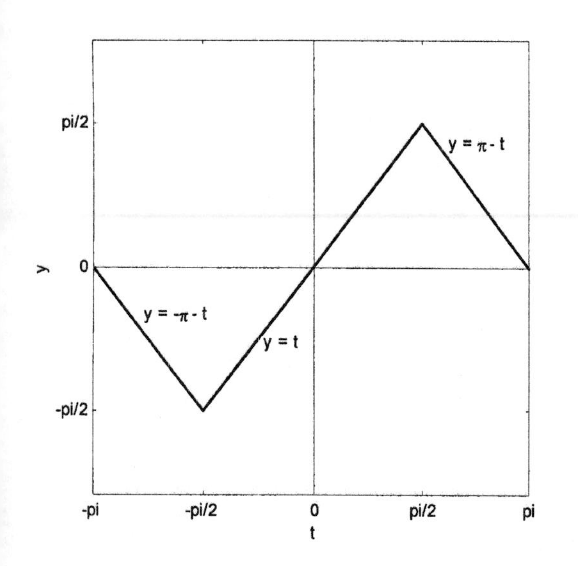

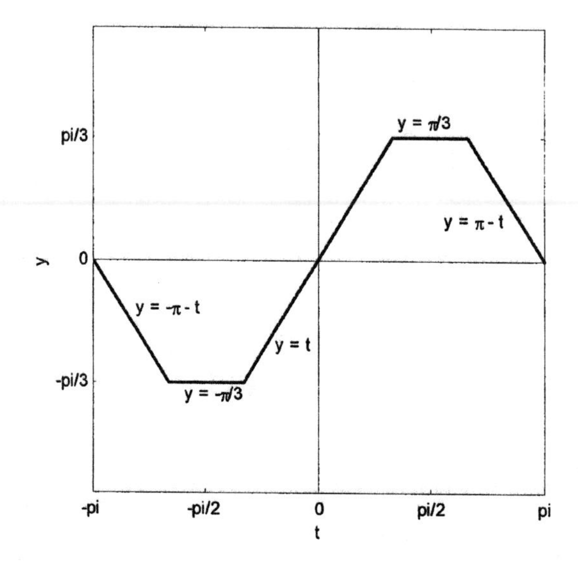

Using *Maple*

We can define the cosine coefficients in (3) as functions of n by

```
a := n -> (1/Pi)*(int(-cos(n*t), t=-Pi..0) +
                    int(+cos(n*t), t=0..Pi)):

a(n);
```

$$0$$

Of course, our odd function has no cosine terms in its Fourier series. The sine coefficients are defined by

```
b := n -> (1/Pi)*(int(-sin(n*t), t=-Pi..0) +
                    int(+sin(n*t), t= 0..Pi)):

b(n);
```

$$-2\frac{-1+\cos(\pi n)}{\pi n}$$

Then a typical partial sum of the Fourier (sine) series is given by

```
fourierSum := sum('b(n)*sin(n*t)', 'n'=1..9);
```

$$fourierSum := 4\frac{\sin(t)}{\pi}+\frac{4}{3}\frac{\sin(3\,t)}{\pi}+\frac{4}{5}\frac{\sin(5\,t)}{\pi}+\frac{4}{7}\frac{\sin(7\,t)}{\pi}+\frac{4}{9}\frac{\sin(9\,t)}{\pi}$$

and we can proceed to plot its graph.

```
plot(fourierSum, t=-2*Pi..4*Pi);
```

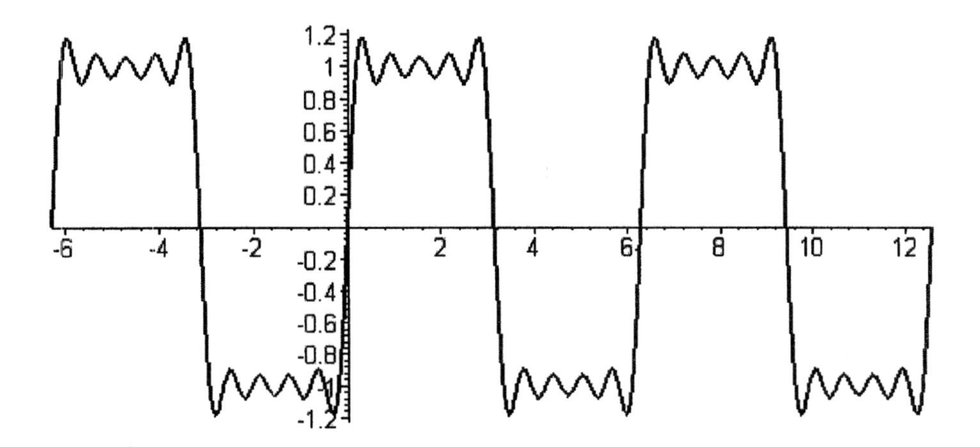

Using *Mathematica*

We can define the cosine coefficients in (3) as functions of n by

```
a[n_] = (1/Pi)*(Integrate[-Cos[n*t], {t,-Pi, 0}] +
                Integrate[+Cos[n*t], {t, 0, Pi}])
```

$$0$$

Of course, our odd function has no cosine terms in its Fourier series. The sine coefficients are defined by

```
b[n_] = (1/Pi)*(Integrate[-Sin[n*t], {t,-Pi, 0}] +
                Integrate[+Sin[n*t], {t, 0, +Pi}]);

b[n] // Simplify
```

$$-\frac{2(\cos(n\pi)-1)}{n\pi}$$

Then a typical partial sum of the Fourier (sine) series is given by

```
fourierSum = Sum[b[n] Sin[n t], {n,1,19}]
```

$$\frac{4\sin(t)}{\pi}+\frac{4\sin(3t)}{3\pi}+\frac{4\sin(5t)}{5\pi}+\frac{4\sin(7t)}{7\pi}+\frac{4\sin(9t)}{9\pi}+$$
$$\frac{4\sin(11t)}{11\pi}+\frac{4\sin(13t)}{13\pi}+\frac{4\sin(15t)}{15\pi}+\frac{4\sin(17t)}{17\pi}+\frac{4\sin(19t)}{19\pi}$$

and we can proceed to plot its graph.

```
Plot[fourierSum, {t,-2 Pi,4 Pi}];
```

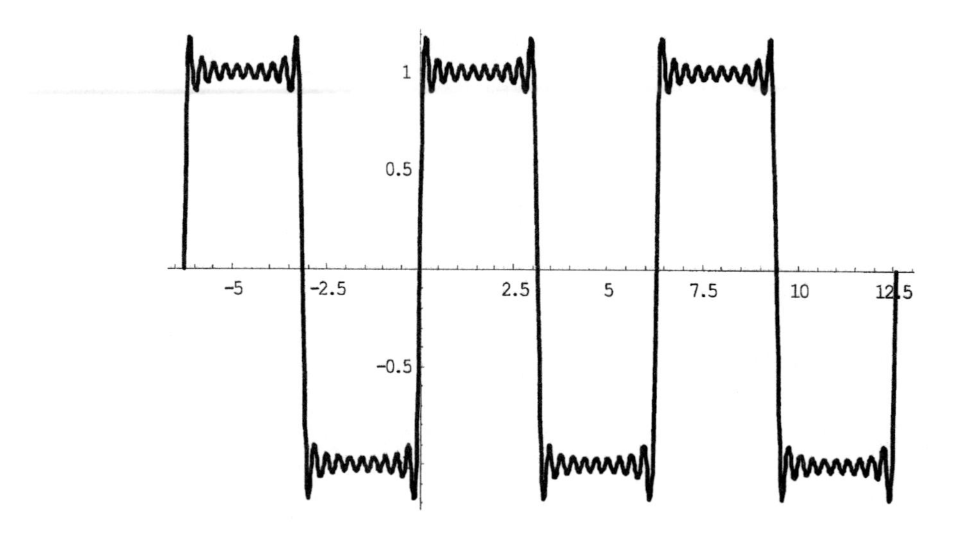

Using MATLAB

We can define the cosine coefficients in (3) as functions of n by

```
syms n t pi
an = (1/pi)*(int(-cos(n*t),-pi,0)+int(cos(n*t),0,pi))
an =
       0
```

Of course, our odd function has no cosine terms in its Fourier series. The sine coefficients are defined by

```
bn = (1/pi)*(int(-sin(n*t),-pi,0)+int(sin(n*t),0,pi));
pretty(bn)
                              -1 + cos(pi n)
                           -2 --------------
                                   pi n
```

MATLAB does not yet know that n is an integer, but *we* do:

```
bn = subs(bn,'(-1)^n','cos(pi*n)');
pretty(bn)
                                         n
                                -1 + (-1)
                             -2 ----------
                                   pi n
```

So now it's obvious that

$$b_n = \begin{cases} 4/n\pi & \text{for } n \text{ odd,} \\ 0 & \text{for } n \text{ even.} \end{cases}$$

We proceed to set up a typical Fourier sum,

```
FourierSum = (4/pi)*sin(t);
for k = 3:2:25
   FourierSum = FourierSum+subs((4/pi)*sin(n*t)/n,k,n);
   end
FourierSum
FourierSum =
4/pi*sin(t)+4/3/pi*sin(3*t)+4/5/pi*sin(5*t)+
4/7/pi*sin(7*t)+ 4/9/pi*sin(9*t)+4/11/pi*sin(11*t)+
4/13/pi*sin(13*t)+4/15/pi*sin(15*t)+4/17/pi*sin(17*t)+
4/19/pi*sin(19*t)+4/21/pi*sin(21*t)+4/23/pi*sin(23*t)+
4/25/pi*sin(25*t)
```

and then to plot its graph.

```
ezplot(FourierSum, 3.1416*[-2  4])
```

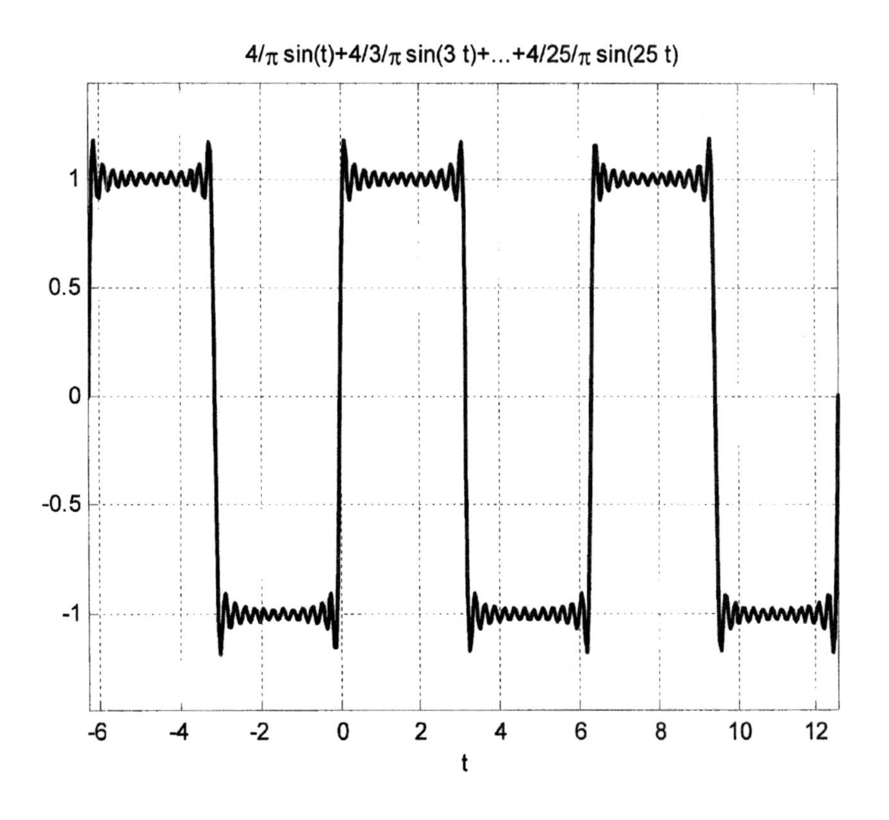

Application 9.3
Fourier Series of Piecewise Smooth Functions

Some computer algebra systems permit the use of unit step functions for the efficient derivation of Fourier series of "piecewise-defined" functions. Let the "unit function" $unit(t, a, b)$ have the value 1 on the interval $a \le t < b$ and the value 0 otherwise. Then we can define a given piecewise smooth function $f(t)$ as a "linear combination" of different unit functions corresponding to the separate intervals on which the function is smooth, with the unit function for each interval multiplied by the formula defining $f(t)$ on that interval. For example, consider the even period 2π function whose graph is shown below. This "trapezoidal wave function" is defined for $0 < t < \pi$ by

$$f(t) = \left(\tfrac{\pi}{3}\right) unit\left(t, 0, \tfrac{\pi}{6}\right) + \left(\tfrac{\pi}{2} - t\right) unit\left(t, \tfrac{\pi}{6}, \tfrac{5\pi}{6}\right) + \left(-\tfrac{\pi}{3}\right) unit\left(t, \tfrac{5\pi}{6}, \pi\right). \tag{1}$$

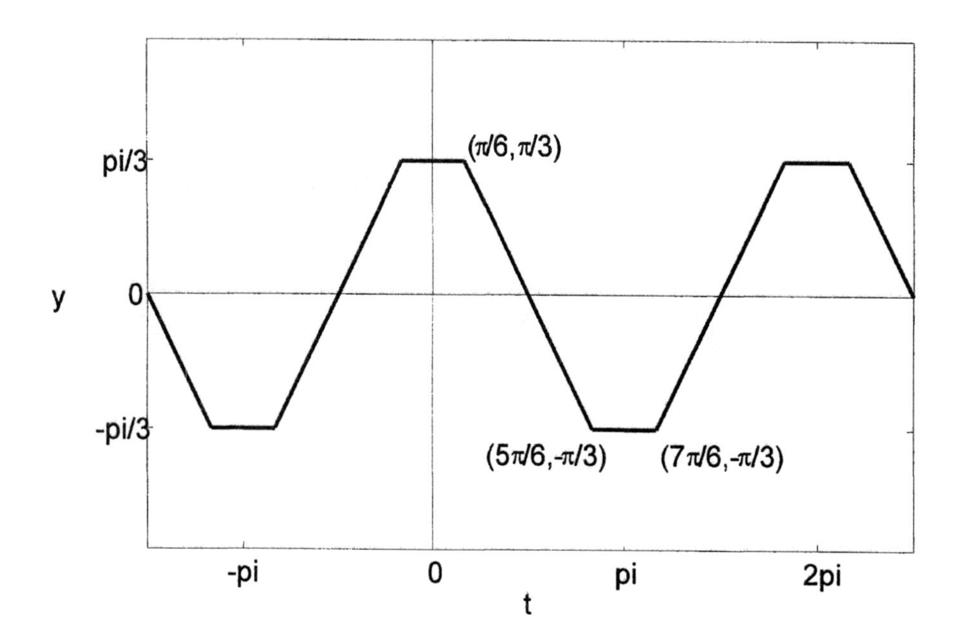

We can then substitute (1) in the Fourier cosine coefficient formula

$$a_n = \frac{2}{\pi} \int_0^{\pi} f(t) \cos nt \, dt \tag{2}$$

to discover the lovely Fourier series

$$f(t) = \frac{2\sqrt{3}}{\pi} \sum \frac{(\pm) \cos nt}{n^2} \tag{3}$$

with a $+ - - + + - - + + \ldots$ pattern of signs, the summation being taken over all odd positive integers n that are *not* multiples of 3.

In the paragraphs that follow we illustrate this approach using *Maple*, *Mathematica*, and MATLAB. You can then apply this method to find the Fourier series of the following period 2π functions:

1. The even square wave function whose graph is shown in Figure 9.3.8 in the text.

2. The even and odd triangular wave functions whose graphs are shown in Figures 9.2.4 and 9.3.9.

3. The odd trapezoidal wave function whose graph in shown in Figure 9.2.5.

Then find similarly the Fourier series of some piecewise smooth functions of your own choice, perhaps ones that have periods other than 2π and are neither even nor odd.

Using *Maple*

The unit step function (with values 0 for $t < 0$ and 1 for $t > 0$) is available in *Maple* as the "Heaviside function":

```
[Heaviside(-2), Heaviside(3)];
```

$$[0, 1]$$

The "unit function" on the t-interval $[a, b]$ is then defined by

```
unit :=
(t,a,b) -> Heaviside(t-a)  - Heaviside(t-b):
```

You can

```
plot(unit(t,1,2), t=0..3);
```

to verify that this unit function looks as intended. The trapezoidal wave function in (1) is then defined for $0 \le t \le \pi$ by

```
f := t ->     (Pi/3)*unit(t,    0,    Pi/6)+
            (Pi/2 - t)*unit(t, Pi/6,  5*Pi/6) +
               (-Pi/3)*unit(t, 5*Pi/6,  Pi):
```

and the graph resulting from the command

```
eps := 10^(-10):
```

```
plot(f(t), t=0+eps..Pi-eps, thickness=2);
```

(from $t \approx 0$ to $t \approx \pi$) agrees with the previous figure. The Fourier cosine coefficients of the even period 2π extension of f are defined by

```
L := Pi:
a := n -> (2/L)*int(f(t)*cos(n*Pi*t/L), t=0..L):
```

The a typical partial sum of the Fourier cosine series of $f(t)$ is given by

```
fourierSum :=
 a(0)/2 + sum(a(n)*cos(n*Pi*t/L), n=1..25);
```

$$fourierSum := 2\frac{\sqrt{3}\cos(t)}{\pi} - \frac{2}{25}\frac{\sqrt{3}\cos(5\,t)}{\pi} - \frac{2}{49}\frac{\sqrt{3}\cos(7\,t)}{\pi} + \frac{2}{121}\frac{\sqrt{3}\cos(11\,t)}{\pi} + \frac{2}{169}\frac{\sqrt{3}\cos(13\,t)}{\pi} - \frac{2}{289}\frac{\sqrt{3}\cos(17\,t)}{\pi} - \frac{2}{361}\frac{\sqrt{3}\cos(19\,t)}{\pi} + \frac{2}{529}\frac{\sqrt{3}\cos(23\,t)}{\pi} + \frac{2}{625}\frac{\sqrt{3}\cos(25\,t)}{\pi}$$

in agreement with (3). You can graph this Fourier sum,

```
plot(fourierSum, t=-3*Pi/2..5*Pi/2);
```

to verify that Fourier series in (3) is consistent with the trapezoidal wave figure shown previously.

Using *Mathematica*

The unit step function (with values 0 for $t < 0$ and 1 for $t > 0$) is available for *Mathematica* calculations if we load the delta functions package:

```
Needs["Calculus`DiracDelta`"]
```

The "unit function" on the t-interval $[a, b]$ is then defined by

```
unit[t_,a_,b_] := UnitStep[t-a] - UnitStep[t-b]
```

You can

```
Plot[unit[t,1,2], {t,0,3}]
```

to verify that this unit function looks as intended. The trapezoidal wave function in (1) is then defined for $0 \le t \le \pi$ by

```
f[t_] :=       (Pi/3) unit[t, 0,       Pi/6] +
           (Pi/2 - t) unit[t, Pi/6, 5*Pi/6] +
              (-Pi/3) unit[t, 5*Pi/6,    Pi]
```

It is clear that $a_0 = 0$ (why?), as we verify with the computation

```
L    = Pi;
a[0]   = (2/L)*Integrate[f[t], {t,0,L}]
```

0

The coefficient a_n is given for $n > 0$ by

```
a[n_]  := (2/L)*Integrate[f[t] Cos[n*Pi*t/L], {t,0,L}]
```

and finally a typical partial sum of the Fourier cosine series of $f(t)$ is given by

```
fourierSum = a[0]/2 + Sum[a[n]*Cos[n*Pi*t/L],{n,1,25}]
```

$$\frac{2\sqrt{3}\cos(t)}{\pi} - \frac{2\sqrt{3}\cos(5t)}{25\pi} - \frac{2\sqrt{3}\cos(7t)}{49\pi} + \frac{2\sqrt{3}\cos(11t)}{121\pi} + \frac{2\sqrt{3}\cos(13t)}{169\pi} - \frac{2\sqrt{3}\cos(17t)}{289\pi} - \frac{2\sqrt{3}\cos(19t)}{361\pi} + \frac{2\sqrt{3}\cos(23t)}{529\pi} + \frac{2\sqrt{3}\cos(25t)}{625\pi}$$

in agreement with (3). You can graph this Fourier sum,

```
Plot[Evaluate[fourierSum], {t,-3 Pi/2,5 Pi/2}]
```

to verify that Fourier series in (3) is consistent with the trapezoidal wave figure shown previously.

Using MATLAB

The "unit function" on the interval $[a, b]$ can be defined by in MATLAB by the m-file

```
function  y = unit(t,a,b)
y = (sign(t-a)  - sign(t-b))/2;
```

You can enter the commands

```
t = 0 : pi/300 : pi;
plot(t, unit(t,1,2))
```

to verify that this unit function looks as intended. The trapezoidal wave function in (1) is then defined for $0 \le t \le \pi$ by

```
function y = f(t)
y =      (pi/3)*unit(t, 0,pi/6) + ...
     (pi/2-t).*unit(t, pi/6,5*pi/6) + ...
       (-pi/3)*unit(t, 5*pi/6,pi);
```

and the graph resulting from the commands

```
t = 0 : pi/300 : pi;
plot(t, f(t))
```

agrees with the trapezoidal graph shown previously.

For the calculation of Fourier coefficients of our trapezoidal function, it is convenient to note that it can be defined by

$$f(t) = \begin{cases} \pi/3 & \text{if } 0 \le t < \pi/6, \\ \pi/2 - t & \text{if } \pi/6 \le t < 5\pi/6, \\ -\pi/3 & \text{if } 5\pi/6 \le t < \pi. \end{cases}$$

Hence we integrate termwise to calculate the Fourier cosine coefficients of the even period 2π extension. First we verify that the constant term in the series is 0.

```
syms t pi
I1 = int(pi/3, t,0,pi/6);
I2 = int(pi/2-t, t,pi/6,5*pi/6);
I3 = int(-pi/3, t,5*pi/6,pi);
a0 = (2/pi)*(I1 + I2 + I3)
a0 =
    0
```

Then we use a loop to calculate and add successive terms one at a time, in order to calculate a partial sum of the series.

```
fourierSum = a0
for n = 1 : 30
     I1 = int((pi/3)*cos(n*t), t,0,pi/6);
     I2 = int((pi/2-t)*cos(n*t), t,pi/6,5*pi/6);
     I3 = int((-pi/3)*cos(n*t), t,5*pi/6,pi);
     an = I1 + I2 + I3;
     term = an*cos(n*t);
     fourierSum = fourierSum + term;
     end
```

Then the desired Fourier sum of our even period 2π trapezoidal wave function is given by

```
fourierSum = (2/pi)*fourierSum;
pretty(fourierSum,60)
```

```
     1/2                   1/2                    1/2
2 (3    cos(t) - 1/25 3    cos(5 t) - 1/49 3    cos(7 t)

             1/2                    1/2                    1/2
   + 1/121 3    cos(11 t) + 1/169 3    cos(13 t) - 1/289 3    cos(17 t)

             1/2                    1/2                    1/2
   - 1/361 3    cos(19 t) + 1/529 3    cos(23 t) + 1/625 3    cos(25 t)

                 1/2
        - 1/841 3    cos(29 t)) / pi
```

in agreement with (3). You can enter the command

```
ezplot(fourierSum, [-5  8])
```

to verify that this Fourier series is consistent with the trapezoidal wave figure shown previously.

Application 9.5
Heated Rod Investigations

In the technology-specific sections below we illustrate the use of *Maple*, *Mathematica*, and MATLAB to investigate numerically the temperature function

$$u(x,t) = \frac{4T_0}{\pi} \sum_{n\,odd} \frac{1}{n} \exp\left(-\frac{n^2\pi^2 kt}{L^2}\right) \sin\frac{n\pi x}{L} \tag{1}$$

of the heated rod of Example 2 in Section 9.5 of the text. This rod of length $L = 50$ cm and thermal diffusivity $k = 0.15$ (for iron) has constant initial temperature $u(x, 0) = T_0 = 100°$ and zero endpoint temperatures $u(0, t) = u(L, t) = 0$ for $t > 0$.

Once we have defined a function that calculates partial sums of the series in (1), we can graph $u(x, t)$ in order to investigate the rate at which the initially heated rod cools. As a practical matter, $N = 50$ terms suffice to give the value $u(x, t)$ after 10 seconds (or longer) with two decimal places of accuracy throughout the interval $0 \le x \le 50$. (How might you check this assertion?)

The graph of $u(x, 1800)$ in the figure below shows the rod's interior temperatures after 30 minutes, and illustrates the fact (?) that the rod's maximum temperature is always at its midpoint, where $x = 25$ cm.

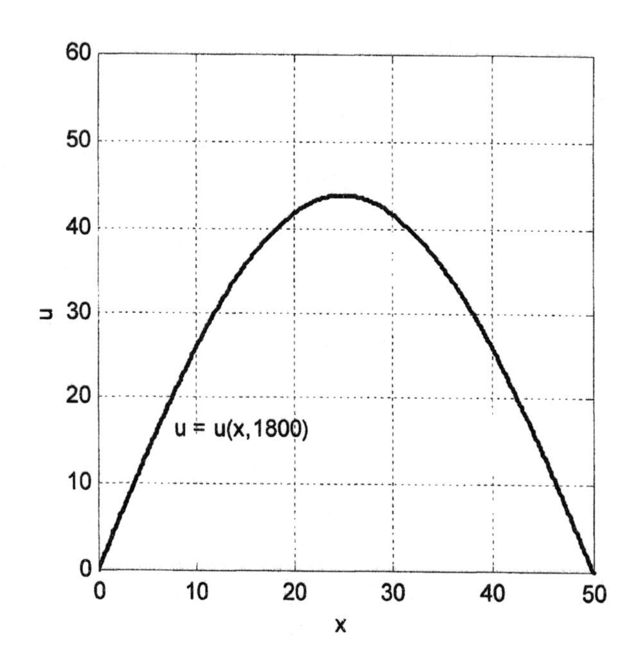

The next graph — of $u(25, t)$ for an initial 50-minute period — indicates that the midpoint ($x = 25$) temperature takes something more than 1500 seconds (25 minutes) to fall to 50°. An appropriate magnification of this graph near its intersection point with the horizontal line $u = 50$ indicates that this actually takes about 1578 seconds (26 min 18 sec).

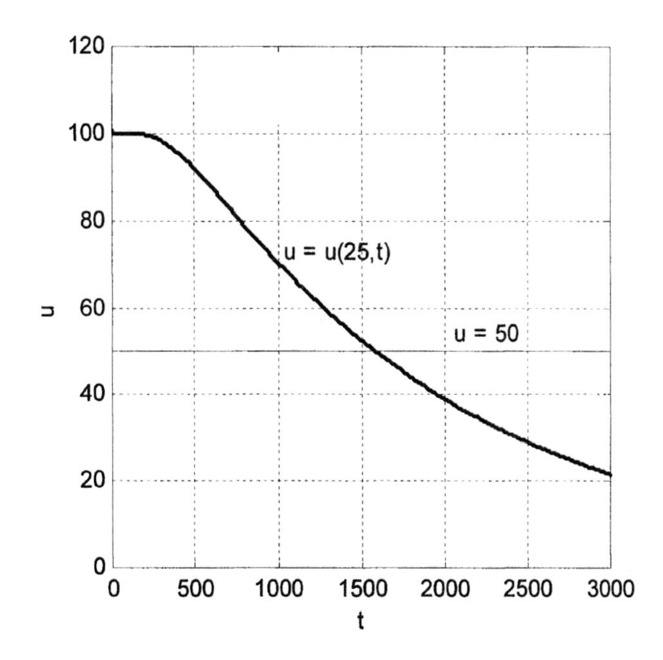

For your personal rod with constant initial temperature $u(x, 0) = T_0 = 100°$ to investigate in this manner, let

$$L = 100 + 10p \quad \text{and} \quad k = 1 + (0.1)q$$

where p is the largest and q is the smallest nonzero digit of your student ID number.

Investigation 1
Investigate more generally the evolution of temperatures in the rod. Show graphs of u versus x for selected values of t. Then determine how long it will be until the maximum temperature anywhere in the rod is 50 degrees You might begin graphically with an appropriate graph of u versus t.

Investigation 2
If the end $x = 0$ of the rod is held at temperature 0, while the end $x = L$ is insulated, then Problem 24 in Section 9.5 says that its temperature function $u(x,t)$ is given by

$$u(x,t) = \sum_{n \text{ odd}}^{\infty} c_n \exp\left(-\frac{n^2\pi^2 kt}{4L^2}\right)\sin\left(\frac{n\pi x}{2L}\right) \tag{2}$$

where

$$c_n = \frac{2}{L}\int_0^L f(x)\sin\frac{n\pi x}{2L}\,dx \tag{3}$$

for $n = 1, 3, 5, \ldots$. Investigate generally the evolution of temperatures in the rod. Show graphs of u versus x for selected values of t. Then determine how long it will be until the maximum temperature anywhere in the rod is 50 degrees.

Using *Maple*

The following *Maple* function sums the first m nonzero terms of the series in (1).

```
L   := 50:       # rod's length
k   := 0.15:     # its diffusivity
T0  := 100:      # initial temperature
m   := 50:       # number of terms

u := (x,t) ->
 (4*T0/Pi)*sum('exp(-(2*n-1)^2*Pi*Pi*k*t/L^2)*
                sin((2*n-1)*Pi*x/L)/(2*n-1)',
               'n' = 1..m ):
```

For instance, the midpoint temperature

```
evalf(u(25,1800));
```

43.848977

after 30 minutes agrees with that found in Example 2 of the text. The commands

```
plot(u(x,1800), x = 0..50);

plot(u(25,t), t = 0..3600);
```

produce graphs like those shown previously. Finally, the command

```
fsolve(u(25,t)=50, t=1500..1600);
```

1578.115993

tells us how long it takes the rod's midpoint temperature to fall to 50°.

Using *Mathematica*

The following *Mathematica* function sums the first m nonzero terms of the series in (1).

```
L  = 50;          (* rod's length          *)
k  = 0.15;        (* its diffusivity       *)
T0 = 100;         (* initial temperature   *)
m  = 50;          (* number of terms       *)

u[x_,t_] :=
(4*T0/Pi)*Sum[Exp[-n*n*Pi*Pi*k*t/L^2]*Sin[n*Pi*x/L]/n,
              {n, 1, 2*m -1, 2} ] // N
```

For instance, the midpoint temperature

```
u[25,1800]

43.849
```

after 30 minutes agrees with that found in Example 2 of the text. The commands

```
Plot[u[x,1800], {x, 0,50}];
```

and

```
Plot[u[25,t], {t, 0,7200}];
```

produce graphs like those shown previously. Finally, the command

```
FindRoot[u[25,t] == 50, {t,1500,1600}]

{t ->1578.12}
```

tells us how long it takes the rod's midpoint temperature to fall to 50°.

Using MATLAB

The following MATLAB function sums the first m nonzero terms of the series in (1).

```
function  u = u(x,t)

k = 0.15;       % diffusivity
L = 50;         % length of rod
u0 = 100;       % initial temperature
S = 0;          % initial sum
N = 50;         % number of terms
for  n = 1:2:2*N+1;
   S = S + ...
       (1/n)*exp(-n^2*pi^2*k*t/L^2).*sin(n*pi*x/L);
   end
u = 4*u0*S/pi;
```

For instance, the midpoint temperature

```
u(25,1800)
ans =
   43.8490
```

after 30 minutes agrees with that found in Example 2 of the text. The commands

```
x = 0 : 0.1 : 50;
U = u(x,1800);
plot(x, U)
```

and

```
t = 0 : 10 : 3600;
U = u(25,t);
plot(t, U)
```

produce graphs like those shown previously. Finally, to find how long it takes the rod's midpoint temperature to fall to 50°, we first define the inline function

```
f = inline('u(25,t)-50')
f =
   Inline function:
   f(t) = u(25,t)-50
```

that gives the difference after t seconds between the midpoint temperature and 50°. Then the result

```
fzero(f, [1500 1600])
Zero found in the interval: [1500, 1600]
ans =
     1.5781e+003
```

shows that this takes about 1578 seconds = 26 min 18 sec.

Further Investigation

We can use *Mathematica* to gain further insight into the temperature function $u(x,t)$; *Maple* and MATLAB users can substitute appropriate commands for those shown below. First, the following command will create a three-dimensional graph of $u(x,t)$; the result should be very similar to Figure 9.5.4 in the text:

```
Plot3D[u[x, t], {x, 0, 50}, {t, 0, 1500},
     ImageSize -> 6*72, ViewPoint -> {2.4,2.1,0.75}];
```

Now we ask this question:

> What if the constant *k*—the thermal diffusivity—were to vary? What change would we expect in $u(x,t)$?

We recall that the "thermal diffusivity" of a medium is a measure of how rapidly heat moves through that medium. If *k* were *less* than the current value of 0.15, then how would you expect the graph of $u(x,t)$ to look different? What if *k* were *greater* than 0.15?

The following *Mathematica* code will produce an animation of the graph of $u(x,t)$ for values of *k* ranging from 0.01 to 0.15 in increments of 0.01. (This assumes that you have executed the commands above that define $u(x,t)$.) Double-click on any of the frames to play the animation; during play you can use the number keys 1,2,3,... to adjust the speed of play:

```
Table[ Plot3D[u[x, t], {x, 0, 50}, {t, 0, 1500},
       ImageSize -> 6*72, ViewPoint -> {3,1.5,0.6}],
       {k, 0.01, 0.15, 0.01}];
```

What happens as *k* increases? Is this consistent with what you predicted above?

Yet another way to illustrate the cooling of this rod is through the use of color. The following **Do** loop will generate an contour map animation that portrays the "hot" areas of the rod in red and the "cool" areas in blue as *t* increases from 0 to 1500 in increments of 50:

```
Do[  temp[x_, y_] = u[x, t];
     ContourPlot[temp[x, y], {x,0,50}, {y,0,10},
     AspectRatio -> Automatic, ImageSize -> 9*72,
     ContourLines -> False,  Contours -> 150,
     ColorFunction -> (Hue[0.7(1 - #/100)]&),
     FrameTicks -> {Automatic, None},
     ColorFunctionScaling -> False],
  {t, 0, 1500,50}];
```

Does this animation give the same impression as the three-dimensional plot of $u(x,t)$ drawn earlier?

Application 9.6
Vibrating String Investigations

The d'Alembert solution

$$y(x,t) = \tfrac{1}{2}[F(x+at)+F(x-at)] \tag{1}$$

of the vibrating string problem (with fixed endpoints and zero initial velocity on the interval $[0, L]$) is readily implemented using a computer algebra system such as *Maple*, *Mathematica*, or MATLAB. Recall that $F(x)$ in (1) denotes the *odd* period 2L extension of the string's initial position function $f(x)$. A plot of $y = y(x,t)$ for $0 \le x \le L$ with t fixed shows a snapshot of the string's position at time t.

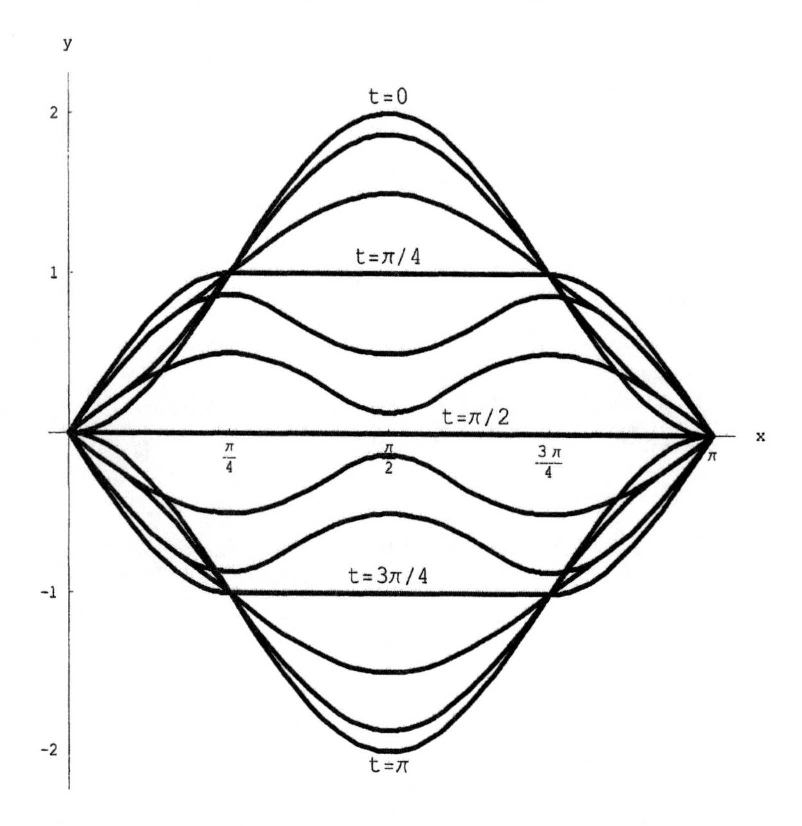

In the paragraphs that follow we illustrate the use of *Maple*, *Mathematica*, and MATLAB to plot such snapshots for the vibrating string with initial position function

$$f(x) = 2\sin^2 x, \qquad 0 \le x \le \pi. \tag{2}$$

We can illustrate the motion of this vibrating string by plotting a sequence of snapshots, either separately (as in Fig. 9.6.4 of the text) or on a single figure — as in the preceding figure. The apparent "flat spots" in the $t = \pi/4$ and $t = 3\pi/4$ snapshots are discussed in Problem 23 of Section 9.6 in the text.

You can test your implementation of d'Alembert's method by attempting to generate Figures 9.6.6 (for a string with triangular initial position) and 9.6.7 (for a string with trapezoidal initial position) in the text. The initial position "bump function"

$$f(x) = \sin^{200} x, \qquad 0 \le x \le \pi. \tag{3}$$

generates travelling waves traveling (initially) in opposite directions, as indicated in Fig. 9.6.3 in the text. The initial position function defined by

$$f(x) = \begin{cases} \sin^{200}(x+1.5) & \text{for } 0 < x < \pi/2, \\ 0 & \text{for } \pi/2 < x < \pi. \end{cases} \tag{4}$$

generates a single wave that starts at $x = 0$ and (initially) travels to the right. (Think of a jump rope tied to a tree, whose free end is initially "snapped".)

After exploring some of the possibilities indicated above, try some initial position functions of you own choice. Any continuous function f such that $f(0) = f(L) = 0$ is fair game. The more exotic the resulting vibration of the string, the better.

Using *Maple*

To plot the snapshots shown simultaneously in the preceding figure, we start with the string's initial position function $f(x)$ defined by

```
f := x -> 2*sin(x)^2:
```

To define the odd period 2π extension $F(x)$ of $f(x)$, we need the following function $s(x)$ that shifts the point x by a multiple of π into the interval $[-\pi, \pi]$.

```
s := proc(x)
     local k;
     k := floor(evalf(x/Pi)):
     if type(k, even)
        then evalf(x - k*Pi):
        else evalf(x - k*Pi - Pi) fi
     end:
```

Then the desired odd extension is defined by

```
F := proc(x)
     if s(x) > 0 then f(s(x)) else -f(s(-x)) fi
     end:
```

Finally, the d'Alembert solution in (1) is defined by

```
G := (x,t) -> ( F(x+t) + F(x-t) )/2:
```

The command

```
plot('G(x,Pi/4)', x = 0..Pi);
```

now gives the $t = \pi/4$ snapshot exhibiting the apparent flat spot previously mentioned. (The quotes are used to prevent premature evaluation during plotting.)

In order to plot the simultaneously the graphs for

$$t = 0, \pi/12, \pi/6, \pi/4, \pi/4, 5\pi/12, \cdots, \pi,$$

we first define the snapshot showing the string's position at time $t = n\pi/12$ by means of the function

```
fig := n -> plot('G(x,n*Pi/12)', x = 0..Pi):
```

Then the preceding figure, exhibiting simultaneously the successive positions of the string in a single composite figure, is generated by the command

```
with(plots):
display([seq(fig(n), n = 0..12)]);
```

If we "restart" with the initial position function

```
f := proc(x)
     if x < Pi/2 then x else Pi - x fi
     end:
```

corresponding to the triangular wave function of Application 9.2, then we get in this way the composite picture shown in Fig 9.6.6 of the text.

Similarly, the trapezoidal wave function

```
f := proc(x)
     if x <= Pi/3 then x
        elif  x > Pi/3 and x < 2*Pi/3
          then Pi/3
        else Pi - x  fi
     end:
```

of Application 9.2 produces the picture shown in Fig. 9.6.7.

Using *Mathematica*

To plot the snapshots shown simultaneously in the preceding figure, we start with the string's initial position function $f(x)$ defined by

```
f[x_] := 2 Sin[x]^2
```

To define the odd period 2π extension $F(x)$ of $f(x)$, we need the following function $s(x)$ that shifts the point x by a multiple of π into the interval $[-\pi, \pi]$.

```
s[x_] := Block[{k}, k = Floor[N[x/Pi]];
            If[EvenQ[k], (* k is even *)
            (* then *)   N[x - k*Pi],
            (* else *)   N[x - k*Pi - Pi]] ]
```

Then the desired odd extension of the initial position function is defined by

```
F[x_] := If[s[x] > 0, (* then *)   f[ s[x]],
                      (* else *)  -f[-s[x]] ]
```

Finally, the d'Alembert solution in (1) is

```
G[x_,t_] := (F[x+t] + F[x-t])/2
```

A snapshot of the position of the string at time t is plotted by

```
stringAt[t_] :=
Plot[G[x,t], {x,0,Pi}, PlotRange -> {-2,2}];
```

For example, the command

```
stringAt[Pi/4];
```

plots the $t = \pi/4$ snapshot exhibiting the apparent flat spot previously mentioned.

We can plot the simultaneously the graphs for

$$t = 0, \pi/12, \pi/6, \pi/4, \pi/4, 5\pi/12, \cdots, \pi,$$

by defining a whole sequence of snapshots at once:

```
snapshots = Table[stringAt[t], {t,0,Pi,Pi/12}];
```

These snapshots can be animated to show the vibrating string in motion, or we can exhibit simultaneously the successive positions of the string in a single composite figure (as shown previously) with the command

```
Show[snapshots];
```

The initial position function

```
f[x_] := If[ x < Pi/2, (* then *) x,
                       (* else *) Pi - x ] // N
```

corresponding to the triangular wave function of Application 9.2 generates in this way the composite picture shown in Fig 9.6.6 of the text.

Similarly, the trapezoidal wave function

```
f[x_] := Which[     0 <= x < Pi/3,       x,
                 Pi/3 <= x < 2*Pi/3,   Pi/3,
               2*Pi/3 <= x <= Pi,     Pi - x ] // N
```

of Application 9.2 produces the picture shown in Fig. 9.6.7.

Using MATLAB

To plot the snapshots shown simultaneously in the preceding figure, we start with the string's initial position function $f(x)$ defined by

```
function  y = f(x)
y = 2*sin(x).^2;
```

saved as the file **f.m**.

To define the odd period 2π extension $F(x)$ of $f(x)$, we need first to shift the point x by a multiple of 2π to a point s in the interval $[-\pi, \pi]$. We then define $F(x)$ to be $f(s)$ if $s>0$, $-f(-s)$ if $x<0$. This is accomplished by the function

```
function y = foddext(x)
%  Odd period 2Pi extension of the function f
k = floor(x/pi);
q = ( 2*floor(k/2) ~= k );   % q = 0 if k even
s = x - (k+q)*pi;            % q = 1 if k odd
m = sign(s);
%  if s>0 then y = f(s) else y = -f(-s)
y = m.*f(m.*s);
```

saved in the file **oddext.m**.

The d'Alembert solution in (1) is now defined by

```
function  y = G(x,t)
y = (foddext(x+t) + foddext(x-t))/2;
```

Then the commands

```
x = 0 : pi/300 : pi;
plot(x, G(x,pi/4))
```

plot the $t = \pi/4$ snapshot exhibiting the apparent flat spot previously mentioned. The simple loop

```
for n = 0 : 12
    plot(x, G(x, n*pi/12))
    axis([0 pi -2 2]); hold on
    end
```

finally generates the preceding composite figure that exhibits simultaneously the successive positions of the vibrating string from $t = 0$ to $t = \pi$ by steps of $\pi/12$.

The initial position function

```
function   y = f(x)
y = x.*(x < pi/2) + (pi-x).*(x >= pi/2);
```

corresponding to the triangular wave function of Application 9.2 generates in this way the composite picture shown in Fig 9.6.6 of the text.

Similarly, the trapezoidal wave function

```
function  y = f(x)
y = x.*(x <= pi/3) ...
    + (pi/3)*(x > pi/3 & x < 2*pi/3)...
    + (pi - x).*(x >= 2*pi/3);
```

of Application 9.2 produces the picture shown in Fig. 9.6.7.

Chapter 10

Eigenvalues and Boundary Value Problems

Application 10.1
Numerical Eigenfunction Expansions

According to Example 7 in Section 10.1 of the text, the eigenfunction series

$$1 = \sum_{n=1}^{\infty} \frac{2(1-\cos\beta_n)}{\beta_n(1+2\cos^2\beta_n)} \sin\beta_n x \qquad (1)$$

represents the function $f(x) \equiv 1$ (for $0 < x < 1$) in terms of the eigenfunctions $\{\sin\beta_n\}_1^{\infty}$ of the Sturm-Liouville problem

$$\begin{aligned} y'' + \lambda y &= 0 \qquad (0 < x < 1) \\ y(0) = 0, &\qquad y(1) + 2y'(1) = 0 \end{aligned} \qquad (2)$$

(We have taken $A = 1$ in Eq. (32) of Section 10.1.) Here we outline *Maple*-, *Mathematica*-, and MATLAB-based numerical investigations of the expansion in (1).

We saw in the solution to Example 7 in Section 10.1 that the values $\{\beta_n\}_1^{\infty}$ in Eq. (1) are the positive solutions of the equation $\tan x = -2x$. Just as indicated in Fig. 10.1.1 there (for the similar equation $\tan x = -x$), the value of β_n is (for n large) just slightly larger than $(2n-1)\pi/2$. However, because of the discontinuities in the tangent function near these solutions, it is best for automatic root-finding purposes to rewrite our eigenvalue equation in the form

$$\sin x + 2x\cos x = 0. \qquad (3)$$

If we wish to approximate the series in (1) by means of an N-term partial sum, our task is twofold:

- First we must approximate the first N positive solutions $\{\beta_n\}_1^{N}$ of Eq. (1); and
- Then we must use these values effectively to calculate the desired partial sum.

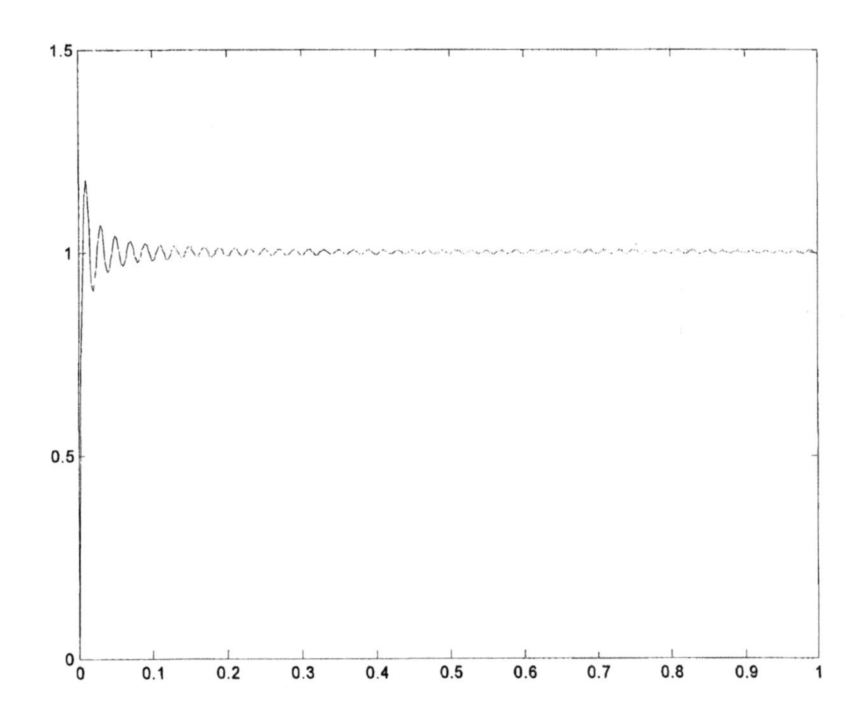

You can check your understanding of eigenfunction expansions by reworking Example 7 in Section 10.1 of the text numerically using an available computing environment, except on the interval $0 < x < L$ and with the right-endpoint condition in (2) replaced with the condition $y(L) + k\,y'(L) = 0$. Select your own values $L > 1$ and $k > 2$. Then calculate the resulting β_n-values and construct a graph like the one above, exhibiting a Gibbs phenomenon at the endpoint $x = 0$.

Using *Maple*

We can solve the equation

```
eq :=    sin(x)+2*x*cos(x)  = 0:
```

for the first

```
N := 100:
```

eigenvalues $\{\beta_n\}_1^N$ by using **`fsolve`** with the nth estimate being slightly larger than $(2n-1)\pi/2$:

```
beta := array(1..N):
for n from 1 to N do
    beta[n]  := fsolve(eq, x=(2*n-1)*Pi/2+0.1):
    od:
```

For instance, the first 10 eigenvalues we get,

```
seq(beta[n], n=1..10);
```

```
1.836597203, 4.815842318, 7.917052685, 11.04082982, 14.17243207,
17.30764054, 20.44480347, 23.58314331, 26.72224637, 29.86187240
```

agree with those shown in Fig. 10.1.4 in the text. The coefficients $\{c_n\}_1^N$ in (1) are then calculated by the loop

```
for n from 1 to N do
    b[n]  := beta[n]:
    c[n]  := 2*(1-cos(b[n]))/(b[n]*(1+2*cos(b[n])^2)):
    od:
```

Finally, the sum of the first N terms in (1) is defined by

```
f := x -> sum(c[n]*sin(b[n]*x), n=1..N):
```

If we display the coefficients here, we find that

$$\begin{aligned} f(x) \approx\ & 1.2083\sin\beta_1 x + 0.3646\sin\beta_2 x + 0.2664\sin\beta_3 x + 0.1722\sin\beta_4 x + \\ & \cdots + 0.0066\sin\beta_{97} x + 0.0065\sin\beta_{98} x + 0.0065\sin\beta_{99} x + 0.0064\sin\beta_{100} x \end{aligned}$$

with the numerical coefficients evidently decreasing quite slowly. The graph of this partial sum is plotted by the commands

```
n := 'n':               # to clear n
plot(f(x), x=0..1, numpoints=500);
```

Using *Mathematica*

We can solve the equation

```
eq =   Sin[x] + 2x Cos[x] == 0;
```

for the first

```
M := 100:
```

eigenvalues $\{\beta_n\}_1^M$ by using **FindRoot** with the nth estimate being slightly larger than $(2n-1)\pi/2$:

```
soln =
Table[FindRoot[eq, {x, (2n-1)Pi/2+0.1}], {n,1,M}];
```

```
beta = x /. soln;
```

For instance, the first 10 eigenvalues we get,

```
Take[beta,10]
{1.8366, 4.81584, 7.91705, 11.0408, 14.1724,
17.3076, 20.4448, 23.5831, 26.7222, 29.8619}
```

agree with those shown in Fig. 10.1.4 in the text. The coefficients $\{c_n\}_1^M$ in (1) are then calculated by the (elementwise) computation

```
b = beta;
c = 2(1 - Cos[b])/(b(1 + 2 Cos[b]^2));
```

Finally, the numerical sum of the first M terms in (1) is defined by

```
f[x_] := Apply[Plus, c Sin[b x]] // N
```

If we display the coefficients here, we find that

$$f(x) \approx 1.2083\sin\beta_1 x + 0.3646\sin\beta_2 x + 0.2664\sin\beta_3 x + 0.1722\sin\beta_4 x + \\ \cdots + 0.0066\sin\beta_{97} x + 0.0065\sin\beta_{98} x + 0.0065\sin\beta_{99} x + 0.0064\sin\beta_{100} x$$

with the numerical coefficients evidently decreasing quite slowly. The graph of our partial sum is plotted by the commands

```
Plot[f[x], {x,0,1},
     PlotPoints->500, PlotRange->{0,1.5}];
```

Using MATLAB

A solution of the equation $\sin x + 2x\cos x = 0$ is a zero of the function

```
f = inline('sin(x)+2*x*cos(x)');
```

We can solve for the first

```
N := 100:
```

eigenvalues $\{\beta_n\}_1^N$ by using **fzero** with the *n*th estimate being a bit larger than $(2n-1)\pi/2$:

```
beta = 1 : N;
```

```
for n = 1 : N
    x0 = (2*n-1)*pi/2;
    beta(n) = fzero(f, [x0  x0+1]);
    end
```

For instance, the first 10 approximate eigenvalues we get,

```
reshape(beta(1:10),[2 5])
ans =
    1.8366     7.9171    14.1724    20.4448    26.7222
    4.8158    11.0408    17.3076    23.5831    29.8619
```

agree with those shown in Fig. 10.1.4 in the text.

The coefficients $\{c_n\}_1^N$ in the eigenfunction series (1) are now calculated by the elementwise computation

```
b = beta;
c = 2*(1-cos(b))./(b.*(1+2*cos(b).^2));
```

If we display the first five and the last five coefficients here,

```
[c(1:5); c(N-4:N)]
ans =
    1.2083     0.3646     0.2664     0.1722     0.1457
    0.0067     0.0066     0.0065     0.0065     0.0064
```

we find that

$$\begin{aligned} f(x) \approx\ & 1.2083\sin\beta_1 x + 0.3646\sin\beta_2 x + 0.2664\sin\beta_3 x + 0.1722\sin\beta_4 x + \\ & \cdots + 0.0066\sin\beta_{97} x + 0.0065\sin\beta_{98} x + 0.0065\sin\beta_{99} x + 0.0064\sin\beta_{100} x \end{aligned}$$

with the numerical coefficients evidently decreasing quite slowly. Finally, the commands

```
x = 0 : 1/500 : 1;
y = zeros(size(x));
for n = 1 : N
    y = y + c(n)*sin(b(n)*x);
end
plot(x,y)
axis([0 1 0 1.5])
```

calculate and graph this partial sum.

Applications 10.2

Numerical Heat Flow Investigations

In the paragraphs that follow we illustrate the use of *Maple*, *Mathematica*, and MATLAB to investigate numerically the temperature function

$$u(x,t) = 2u_0hL\sum_{n=1}^{\infty}\frac{1-\cos\beta_n}{\beta_n(hL+\cos^2\beta_n)}\exp\left(-\frac{\beta_n^{\,2}kt}{L^2}\right)\sin\frac{\beta_n x}{L} \tag{1}$$

of the heated slab of Example 1 in Section 10.2 of the text. We suppose that this slab has thickness $L = 50$ cm, uniform initial temperature $u_0 = 100°$, fixed temperature 0° at the slab's left-hand boundary $x = 0$, thermal diffusivity $k = 0.15$ (for iron), and heat transfer coefficient $h = 0.1$ at the right-hand boundary $x = L$.

According to Eq. (8) in the text, the eigenvalues $\{\beta_n\}_1^\infty$ in (1) are the positive solutions of the equation $\tan x = -x/hL$. Figure 10.1.1 in the text exhibits these eigenvalues as the x-coordinates of the points of intersection of $y = x/hL$ and $y = -\tan x$. There we see that if n is large then β_n is slightly larger than $(2n-1)\pi/2$. This observation provides initial estimates with which we can proceed to approximate the eigenvalues numerically (using Newton's method, for instance). Once a sufficient number of the $\{\beta_n\}_1^\infty$ have been found, we can proceed to sum the series (1) numerically — as needed to plot the temperature u either as a function of x or as a function of t.

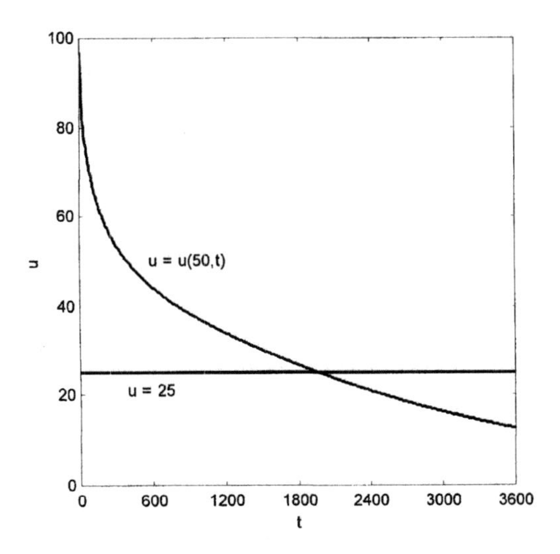

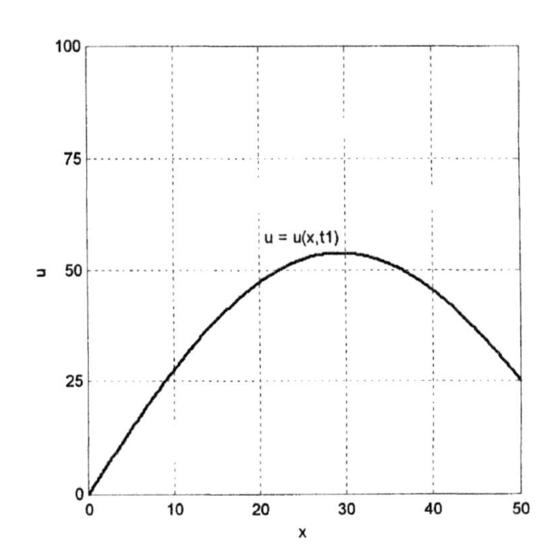

The figure on the left illustrates the falling temperature of the slab at its right-hand boundary where heat transfer with a medium at temperature 0° takes place. This graph of $u = u(50,t)$ for the first hour indicates that the right-hand boundary temperature of the slab falls to 25° in slightly over a half hour (1800 seconds). We will see that this actually takes about $t_1 = 1951$ seconds. The graph $u = u(x,t_1)$ shown in the figure on the right indicates that interior temperatures of the slab at time t_1 still remain above 50° at some points. We will see that the maximum temperature within the slab at this instant is approximately 53.8° and occurs at $x \approx 29.4$.

Investigation A

For your own slab with uniform initial temperature $u(x,0) = 100°$ to investigate in this manner, let $h = 0.1$,

$$L = 5(10 + p) \qquad \text{and} \qquad k = 0.1\,q,$$

where p is the largest and q is the smallest nonzero digit of your student ID number. Then carry out the investigation outlined above to find:

- When the right boundary temperature of the slab is 25°; and
- The maximum temperature within the slab at this instant.

As an alternative to finding roots and minimum values numerically, you can "zoom in" graphically on appropriate points of figures like those shown above.

Investigation B

Here you are to investigate the cooling of a spherical ball of radius a that initially has constant temperature $u_0 = 100^\circ$ throughout. At time $t = 0$ it is immersed in a surrounding medium at temperature 0. At time t the temperature u in the ball then depends only on the distance r from its center, and $u(r,t)$ satisfies the boundary value problem

$$(r\,u)_t = k\,(r\,u)_{xx} \qquad (2)$$

$$H\,u(a,t) + K\,u_r(a,t) = 0 \qquad \text{(heat transfer at boundary)}$$

$$u(r,0) = u_0 \qquad \text{(initial temperature).}$$

The heat transfer condition includes the insulated case $H = 0$ which reduces to $u_r(a,t) = 0$, as well as the zero boundary temperature case $H = \infty$ which (for a simple listing of cases) we take to mean that $u(a,t) = 0$.

To get started, substitute $v(r,t) = r\,u(r,t)$ in (2) to derive the transformed boundary value problem

$$v_t = k\, v_{rr} \qquad (0 < r < a) \qquad (3)$$

$$v(0,t) = 0 \qquad \text{(left endpoint)}$$

$$(H - K/a)\, v(a,t) + K\, v_r(a,t) = 0 \qquad \text{(right endpoint)}$$

$$v(r,0) = r\, u_0$$

involving the standard 1-dimensional heat equation in (3).

Show first that the separation of variables $v(r,t) = R(r)\,T(t)$ leads to an eigenvalue problem of the form

$$y''(x) + \lambda\, y(x) = 0 \qquad (0 < x < L)$$

$$y(0) = 0, \qquad h\, y(L) + y'(L) = 0$$

(in standard x, y, L notation). Solve this problem in the following six cases.

- Case 0: $h = \infty$ (that is, $y(L) = 0$)
- Case 1: $h > 0$
- Case 2: $h = 0$ (that is, $y'(L) = 0$)
- Case 3: $-1 < hL < 0$
- Case 4: $hL = -1$
- Case 5: $hL < -1$

In each case determine *all* the eigenvalues and eigenfunctions (including whether there are any zero or negative eigenvalues). In cases having non-standard eigenvalue equations, draw sketches showing the approximate locations of the eigenvalues, and approximate accurately the first the first several of them. In the remainder of the project you may need the first 40 or 50 eigenvalues for each case, but after a while you can get additional ones simply by adding π successively.

Now for the fun! With this mathematical preparation, you're ready to investigate the cooling of the physical ball. Suppose it's made of a copper alloy with

radius	$a = 10$ cm
density	$\delta = 10$ gm/cm^3
specific heat	$c = 0.1$ cal/gm
conductivity	$K = 1$ (in cgs units)

You are to analyze the following five cases.

- Case 0: $H = \infty$ (that is, $u(a,t) = 0$)
- Case 1: $H = 0.5$
- Case 2: $H = 0.1$
- Case 3: $H = 0.05$
- Case 4: $H = 0$ (that is, $u_r(a,t) = 0$)

In each case, derive an infinite series solution of the form

$$u(r,t) = \sum c_n \exp\left(-\frac{\beta_n^2 kt}{a^2}\right)\frac{R_n(r)}{r}.$$

in terms of the eigenvalue-eigenfunction $\{\beta_n, R_n(t)\}$ pairs for the appropriate eigenvalue problem. Evaluate the appropriate integrals to express c_n in terms of *H*, *a*, and β_n so you can calculate $u(r, t)$ numerically. Produce one or more *ru*-graphs illustrating the cooling within the ball. Finally, use *tu*-graphs to find how long it takes the center of the ball to cool to 50° (except in Case 4 where this makes no sense — explain why).

Using *Maple*

We enter the eigenvalue equation

```
L := 50:            h := 0.1:
eq :=    tan(x)  = -x/(h*L):
```

and proceed to solve it for the first $N = 20$ of the $\{\beta_n\}_1^\infty$ values in (1), as follows:

```
N := 20:
beta := array(1..N):
for n from 1 to N do
    beta[n]  := fsolve(eq, x=(2*n-1)*Pi/2+0.1):
    od:
seq(beta[n], n=1..N);
```

2.653662400, 5.454353755, 8.391345550, 11.40862652, 14.46986802,
17.55621326, 20.65782432, 23.76927763, 26.88739809, 30.01022364,
33.13648417, 36.26532449, 39.39614920, 42.52853172, 45.66215882,
48.79679551, 51.93226220, 55.06841930, 58.20515676, 61.34238666

Note how we make use of the fact that each β_n is slightly larger than $(2n-1)\pi/2$.

Now we can calculate the coefficients $\{c_n\}$ in (1):

```
b := array(1..N):
c := array(1..N):

for n from 1 to N do
    b[n] := beta[n]:
    c[n] := (1-cos(b[n]))/(b[n]*(h*L+cos(b[n])^2)):
    od:
```

The following function then sums the first N nonzero terms of the series in Eq. (1).

```
u0 := 100:      k := 0.15:    n := 'n':
u := (x,t) -> 2*u0*h*L*
                sum( c[n]*exp(-b[n]*b[n]*k*t/L^2)*
                       sin(b[n]*x/L), n = 1..N );
```

As a practical matter, this suffices to calculate the value $u(x,t)$ after 10 seconds or more with 2-place accuracy throughout the interval $0 \le x \le 50$. (How might you verify this assertion?) The command

```
plot({u(50,t), 25}, t=0..3600);
```

yields the preceding graph of $u(50,t)$ for $0 \le t \le 3600$. The solution

```
t1 := fsolve(u(50,t) = 25, t = 1900..2000);
```

$$t1 := 1951.24$$

then shows that it takes just over $t_1 = 1951$ seconds for the right-hand boundary temperature of the slab to fall to 25°. The command

```
plot(u(x,t1), x = 0..50);
```

graphs $u = u(x,t_1)$ for $0 \le x \le 50$, and — observing the maximum near $x = 30$ — we finally maximize the temperature $u(x,t_1)$:

```
maximize(u(x,t1), x, 29..31);
```

$$53.7942$$

Thus the maximum temperature within the slab at time $t = t_1$ is just under 54°.

Using *Mathematica*

We enter the eigenvalue equation

```
L = 50;              h = 0.1;
eq =    Tan[x] == -x/(h*L);
```

and proceed to solve it for the first $M = 20$ of the $\{\beta_n\}_1^\infty$ values in (1), as follows:

```
M = 20;
solutions =
Table[ FindRoot[ eq, {x,(2*n-1)*Pi/2 + 0.1}],
       {n, 1, M}];
beta = x /. solutions

{2.65366, 5.45435, 8.39135, 11.4086, 14.4699,
 17.5562, 20.6578, 23.7693, 26.8874, 30.0102,
 33.1365, 36.2653, 39.3961, 42.5285, 45.6622,
 48.7968, 51.9323, 55.0684, 58.2052, 61.3424}
```

Note how we make use of the fact that each β_n is slightly larger than $(2n-1)\pi/2$. Now we can calculate the coefficients $\{c_n\}$ in (1):

```
b = beta;           k = 0.15;
c = (1 - Cos[b])/(b*(h*L + Cos[b]^2));
```

The following function then sums the first M nonzero terms of the series in Eq. (1).

```
u0 = 100;
u[x_,t_] :=
2*u0*h*L*Apply[ Plus,
                  c*Exp[-b*b*k*t/L^2]*Sin[b*x/L]] // N
```

As a practical matter, this suffices to calculate the value $u(x,t)$ after 10 seconds with two-place accuracy throughout the interval $0 \le x \le 50$. (How might you verify this assertion?) The command

```
Plot[{u[50,t], 25}, {t,0,3600}];
```

yields the preceding graph of $u(50,t)$ for $0 \le t \le 3600$. The solution

```
t1 = t /. FindRoot[u[50,t] == 25, {t,2000}]

1951.24
```

then shows that it takes just over $t_1 = 1951$ seconds for the right-hand boundary temperature of the slab to fall to 25°. The command

```
Plot[u[x,t1], {x,0,50}];
```

graphs $u = u(x,t_1)$ for $0 \le x \le 50$, and — observing the maximum near $x = 30$ — we finally can find the maximum value of the temperature $u(x,t_1)$ by finding the minimum value of its negative $-u(x,t_1)$:

```
FindMinimum[-u[x,t1], {x,30}]
{-53.804, {x -> 29.3586}}
```

Thus the maximum temperature within the slab at time $t = t_1$ is just under 54°.

Using MATLAB

First we define the eigenvalue equation $\tan x = -x/hL$ in the form of the m-file function

```
function  y = f(x)
L = 50;    h = 0.1;
y = tan(x) + x/(h*L);
```

whose successive positive zeros we seek. Then we proceed to find the first $N = 20$ of the $\{\beta_n\}_1^\infty$ values in (1) as follows:

```
N = 20;
beta = 1:N;
for n = 1 : N
beta(n) = fzero('f', [(2*n-1)*pi/2+0.001 n*pi]);
end
reshape(beta,5,4)'
ans =
     2.6537     5.4544     8.3913    11.4086    14.4699
    17.5562    20.6578    23.7693    26.8874    30.0102
    33.1365    36.2653    39.3961    42.5285    45.6622
    48.7968    51.9323    55.0684    58.2052    61.3424
```

Note how we make use of the fact that each β_n is slightly larger than $(2n-1)\pi/2$. Now we can calculate the coefficients $\{c_n\}$ in (1):

```
L = 50;    h = 0.1;
b = beta;
c = (1 - cos(b))./(b.*(h*L + cos(b).^2));
```

As a practical matter, this suffices to calculate the value $u(x,t)$ with two-place accuracy if $0 \le x \le 50$ and $t \ge 10$. (How might you verify this assertion?) For instance, given the slab's initial temperature and thermal diffusivity

```
u0 = 100;        k = 0.15;
```

the calculation

```
x = 50;   t = 3600;
u = 2*u0*h*L*sum(c.*exp(-b.*b*k*t/L^2).*sin(b*x/L))
u =
   12.5627
```

reveals that after 1 hour the slab's right-hand boundary temperature has fallen to about 12.56°.

For plotting purposes it is convenient to assemble the commands above into a function `u(x,t)` that accepts as inputs a q-vector `x` and a p-vector `t` and returns a $p \times q$ matrix `u` of corresponding temperatures:

```
function    u = u(x,t)

%   x = row vector of  q  points of rod
%   t = row vector of  p  times
%   u = p x q  matrix of temperatures

L = 50;                % length of rod
h = 0.1;               % heat transfer coefficient
u0 = 100;              % initial temp of slab
k = 0.15;              % thermal diffusivity
N = 20;                % no of terms to use
for  n = 1 : N        % solve for beta values
     b(n) = fzero('f', [(2*n-1)*pi/2+0.001 n*pi]);
     end

c = 2*u0*h*L*(1 - cos(b))./(b.*(h*L + cos(b).^2));
coeffs = diag(c);                      % N x N  diag matrix
                                       % of coefficients
exps = exp(-(b'.*b')*k*t/L^2);         %  N x p  matrix
sines = sin(b'*x/L);                   %  N x q  matrix
u = exps'*coeffs*sines;                %  p x q  matrix
```

The final line of `u.m` uses matrix multiplication to sum the terms of the series. The commands

```
x = 50;
t = 0 : 10 : 3600;
clear u
U = u(x,t);
plot(t, U), hold on
plot([0 3600], [25 25])
```

then yield the preceding graph of $u(50,t)$ for $0 \le t \le 3600$. If we define the temperature function

```
g = inline('u(50,t)-25','t');
g =
     Inline function:
     g(t) = u(50,t)-25
```

such that $g(t)$ is the difference between 25 and the slab's right-hand boundary temperature at time t, then the solution

```
t1 = fzero(g, [1900 2000])
t1 =
   1.9512e+03
```

shows that it takes just over $t_1 = 1951$ seconds for the right-hand boundary temperature of the slab to fall to 25°. The commands

```
x = 0 : 0.2 : 50;
plot( x, u(x,t1) )
```

graph $u = u(x,t_1)$ for $0 \le x \le 50$, and we spot a maximum near $x = 30$. Finally, we can find the maximum value of the temperature $u(x,t_1)$ by finding the minimum value of its negative $-u(x,t_1)$:

```
h = inline('-u(x,1951.2)','x')
h =
     Inline function:
     h(x) = -u(x,1951.2)

x1 = fmin(h,29,31)
x1 =
   29.3586

u(x1,t1)
ans =
   53.8040
```

Thus the maximum temperature within the slab at time $t = t_1$ is just under 54°.

Application 10.3
Vibrating Beams and Diving Boards

In this project you are to investigate further the vibrations of an elastic bar or beam of length L whose position function $y(x,t)$ satisfies the partial differential equation

$$\rho \frac{\partial^2 y}{\partial t^2} + EI \frac{\partial^4 y}{\partial x^4} = 0 \qquad (0 < x < L) \qquad (1)$$

and the initial conditions $y(x,0) = f(x)$, $y_t(x,0) = 0$.

First separate the variables (as in Example 3 of Section 10.3 in the text) to derive the formal series solution

$$y(x,t) = \sum_{n=1}^{\infty} c_n X_n(x) \cos \frac{\beta_n^2 a^2 t}{L^2} \qquad (2)$$

where $a^4 = EI/\rho$, the $\{c_n\}$ are the appropriate eigenfunction expansion coefficients of the initial position function $f(x)$, and the $\{\beta_n\}$ values and $\{X_n(x)\}$ eigenfunctions are determined by the end conditions imposed on the bar. In a particular case, one wants to find both the **frequency equation** whose positive roots are the $\{\beta_n\}$ and the explicit eigenfunctions $\{X_n(x)\}$. In this section we saw that the frequency equation for the *fixed-fixed* case (with $y(0) = y'(0) = y(L) = y'(L) = 0$) is $\cosh x \cos x = 1$, that is,

$$\operatorname{sech} x = \cos x \qquad (3)$$

The solutions of this equation are located approximately by the graph below, where we see that the nth positive solution is given approximately by $\beta_n \approx (2n+1)\pi/2$.

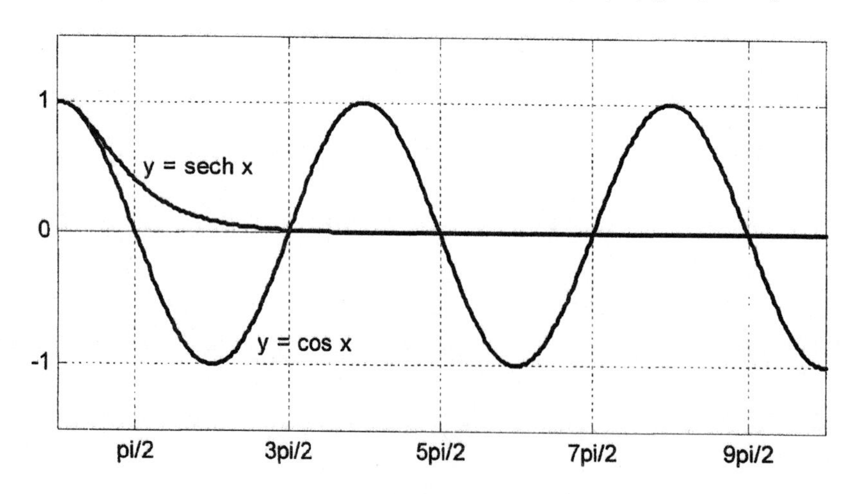

Case 1: Hinged at each end
The endpoint conditions are

$$y(0) = y''(0) = y(L) = y''(L) = 0. \tag{4}$$

According to Problem 8 in Section 10.3 of the text, the frequency equation is $\sin x = 0$, so $\beta_n = n\pi$ and $X_n(x) = \sin n\pi x$ for $n = 1, 2, 3, \ldots$. Suppose that the bar is made of steel (with density δ = 7.75 g/cm^3 and Young's modulus $E = 2 \times 10^{12}$ dyn/cm^2), and is 19 inches long with square cross section of edge $w = 1$ in. (so its moment of inertia is $I = \frac{1}{12}w^4$). Determine its first few natural frequencies of vibration (in Hz). How does this bar sound when it vibrates?

Case 2: Free at each end
The endpoint conditions are

$$y''(0) = y'''(0) = y''(L) = y'''(L) = 0. \tag{5}$$

This case models, for example, a weightless bar suspended in space in an orbiting spacecraft. Now show that the frequency equation is again $\operatorname{sech} x = \cos x$ as in (3), although the eigenfunctions in this *free-free* case differ from those in the fixed-fixed case discussed in the text. From the figure above we see that the *n*th positive solution is given approximately by $\beta_n \approx (2n+1)\pi/2$. Use the numerical methods of Project 10.2 to approximate the first several natural frequencies of free-free vibration of the same physical bar considered in case 1. How does it sound now?

Case 3: Fixed at $x = 0$, free at $x = L$
Now the boundary conditions are

$$y(0) = y'(0) = y''(L) = y'''(L) = 0. \tag{6}$$

This is a **cantilever** like the diving board illustrated in Fig. 10.3.8 of the text. According to Problem 15 there, the frequency equation is

$$\cosh x \cos x = -1. \tag{7}$$

From the figure at the top of the next page, showing the graphs $y = \operatorname{sech} x$ and $y = -\cos x$, we see that $\beta_n \approx (2n-1)\pi/2$ for n large. Use the numerical methods of Project 10.2 to approximate the first several natural frequencies of vibration (in Hz) of the particular diving board described in Problem 15 — it is 4 meters long and is made of steel (with density δ = 7.75 g/cm^3 and Young's modulus $E = 2 \times 10^{12}$ dyn/cm^2); its cross section is a rectangle with width $a = 30$ cm and thickness $b = 2$ cm, so its moment of inertia about a horizontal axis of symmetry is $I = \frac{1}{12}ab^3$. Thus find the frequencies at which a diver on this board should bounce up and down at the free end for maximal resonant effect.

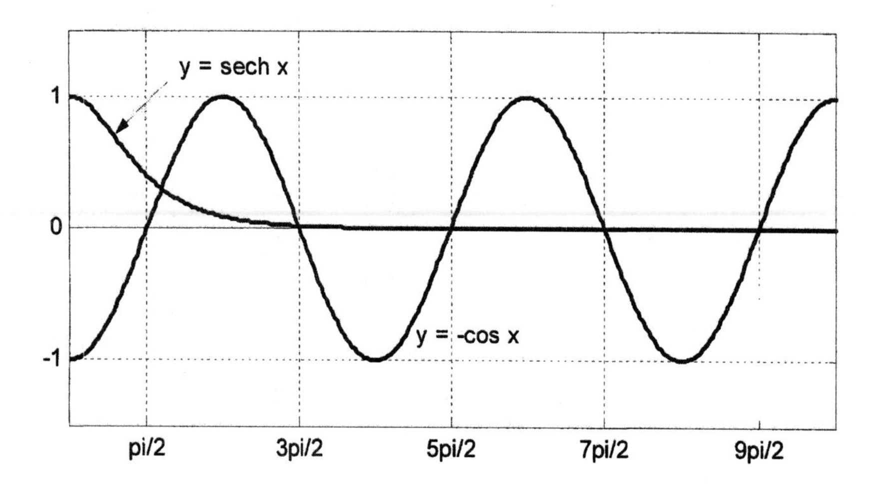

Using *Maple*

We enter the eigenvalue equation

```
eq :=    sech(x)  = cos(x):
```

in (3) and proceed to solve it for the first $N = 10$ of the $\{\beta_n\}_1^\infty$ values, as follows:

```
N := 10:
beta := array(1..N):
for n from 1 to N do
    beta[n] := fsolve(eq, x=(2*n+1)*Pi/2):
    od:
seq(beta[n], n=1..N);
```

4.730040745, 7.853204624, 10.99560784, 14.13716549, 17.27875966,
20.42035225, 23.56194490, 26.70353756, 29.84513021, 32.98672286

Note how we make use of the fact that β_n is approximately $(2n+1)\pi/2$. It is interesting to check successive differences of eigenvalues:

```
seq(beta[n]-beta[n-1], n=2..N);
```

3.123163879, 3.142403216, 3.14155765, 3.14159417, 3.14159259,
3.14159265, 3.14159266, 3.14159265, 3.14159265

We see that, for 6-place accuracy, it suffices to calculate the first half-dozen β_n-values and then add successive multiples of π.

The physical parameters of the fixed-fixed bridge discussed in Section 10.3 of the text are given by

```
delta := 7.75:          # volumetric density
A := 40:                # cross-sectional area
rho := delta*A:         # linear density
L := 120*30.48:         # length of bridge
I0 := 9000:             # moment of inertia
E := 2*10^12:           # Young's modulus
```

Its natural frequencies of vibration, $\omega_n = (\beta_n^2 / L^2)\sqrt{EI_0/\rho}$, are given in radians/sec by

```
b := beta:
w := map(b->(b^2/L^2)*sqrt(E*I0/rho), b):
```

and then in hertz (cycles/min) by

```
map(w->evalf(60*w/(2*Pi)), w);
```

[121.6925898, 335.4503097, 657.6167321, 1087.073862, 1623.900808, 2268.092846, 3019.650238, 3878.572976, 4844.861051, 5918.514469]

Thus the first "critical cadence" for marching soldiers is about 122 steps per minute.

Using *Mathematica*

We enter the eigenvalue equation

```
eq =   Sech[x] == Cos[x];
```

in (3) and proceed to solve it for the first $N = 10$ of the $\{\beta_n\}_1^\infty$ values, as follows:

```
M = 10;
solutions =
Table[ FindRoot[ eq, {x,(2*n+1)*Pi/2}],
       {n, 1, M}];
beta = x /. solutions
```

```
{4.7300, 7.8532, 10.9956, 14.1372, 17.2788,
20.4204, 23.5619, 26.7035, 29.8451, 32.9867}
```

Note how we make use of the fact that β_n is approximately $(2n+1)\pi/2$. It is interesting to check successive differences of eigenvalues:

```
Table[beta[[n]]-beta[[n-1]], {n,2,10}]
```

```
{3.12316, 3.1424, 3.14156, 3.14159,
3.14159, 3.14159, 3.14159, 3.14159, 3.14159}
```

We see that, for 5-place accuracy, it suffices to calculate the first half-dozen β_n-values and then add successive multiples of π.

The physical parameters of the free-free steel bar of case 2 are given by

```
delta = 7.75;           (* volumetric density    *)
A = 2.54^2;             (* cross-sectional area *)
rho = delta*A;          (* linear density        *)
L = 19*2.54;            (* length of bridge      *)
I0 = (2.54^4)/12;       (* moment of inertia     *)
E0 = 2*10^12;           (* Young's modulus       *)
```

Its natural frequencies of vibration, $\omega_n = (\beta_n^2 / L^2)\sqrt{EI_0 / \rho}$, are given in radians/sec by

```
b = beta;
w = (b^2/L^2) Sqrt[E0*I0/rho];
```

and then in cycles per second by

```
w = w/(2 Pi)

{569.485, 1569.81, 3077.45, 5087.18, 7599.37,
10614., 14131.1, 18150.6, 22672.5, 27696.9}
```

Whereas the fundamental frequency of the hinged-hinged bar of case 1 is

```
(Pi^2/L^2) Sqrt[E0*I0/rho]/(2 Pi)

251.219
```

cycles/sec, that is, about middle C, we see that the fundamental frequency of the corresponding free-free bar of case 2 is above high C.

Using MATLAB

First we define the eigenvalue equation $\cosh x \cos x = -1$ in (7), for the fixed-free cantilever of case 3, in the form of the function

```
f = inline('cosh(x)*cos(x)+1','x')
f =
     Inline function:
     f(x) = cosh(x)*cos(x)+1
```

whose successive positive zeros we seek. Then we proceed to find the first $N = 10$ of the $\{\beta_n\}_1^\infty$ values as follows:

```
N = 10;
beta = 1 : N;
for n = 1 : N
    beta(n) = fzero(f, (2*n-1)*pi/2);
    end
reshape(beta,5,2)'
ans =
    1.8751    4.6941    7.8548   10.9955   14.1372
   17.2788   20.4204   23.5619   26.7035   29.8451
```

Note how we make use of the fact that β_n is approximately $(2n-1)\pi/2$.

The physical parameters of the diving board described in Problem 15 in the text are given by

```
delta = 7.75;          % volumetric density
a = 30;                % width of diving board
b = 2;                 % its thickness
A = a*b;               % cross-sectional area
rho = delta*A;         % linear density
L = 400;               % length of diving board
I0 = a*b^3/12;         % moment of inertia
E = 2*10^12;           % Young's modulus
```

Its natural frequencies of vibration, $\omega_n = (\beta_n^2 / L^2)\sqrt{EI_0 / \rho}$, are given in radians/sec by

```
b = beta;
w = (b.^2/L^2)*sqrt(E*I0/rho);
```

and then in cycles per second by

```
w = w/(2*pi)
w =
   1.0258     6.4285    17.9999    35.2725    58.3081
  87.1021   121.6550   161.9668   208.0373   259.8667
```

Thus the diving board's fundamental frequency is about 1.0258 cycles/sec, so for maximal resonant effect the diver should bounce up and down just about once per second.

Application 10.4
Bessel Functions and Heated Cylinders

In the paragraphs that follow we illustrate the use of *Maple*, *Mathematica*, and MATLAB to investigate numerically the temperature function

$$u(r,t) = \sum_{n=1}^{\infty} a_n \exp\left(-\frac{\gamma_n^2 kt}{c^2}\right) J_0\left(\frac{\gamma_n r}{c}\right) \tag{1}$$

of the heated cylindrical rod of Example 1 in Section 10.4 of the text , with radius $c = 10$ cm, having constant initial temperature $u_0 = 100°$ and thermal diffusivity $k = 0.15$ (for iron). The coefficients $\{a_n\}_1^{\infty}$ in (1) depend on the condition imposed at the boundary $r = c$ of the cylinder.

In the case of the zero boundary condition $u(c,t) \equiv 0$, Eq. (30) in Section 10.4 implies that

$$a_n = \frac{2u_0}{\gamma_n J_1(\gamma_n)}, \tag{2}$$

where the $\{\gamma_n\}_1^{\infty}$ are the positive solutions of the equation

$$J_0(x) = 0. \tag{3}$$

Once the first so many of these values have been determined numerically, we can sum the series in (1) so as to calculate values of u and plot $u(r,t)$ versus either r or t. For instance, the graph of $u = u(r,120)$ in the figure on the left below shows how the temperature within the rod after 2 minutes varies with the distance r from its center, and we see that the center-line temperature has already fallen below 60°. The graph of

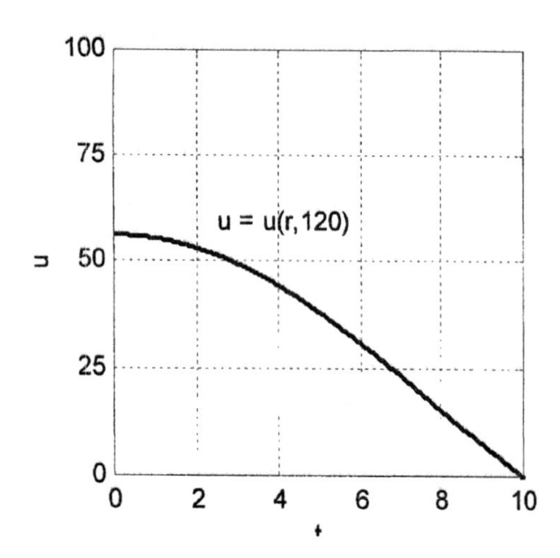

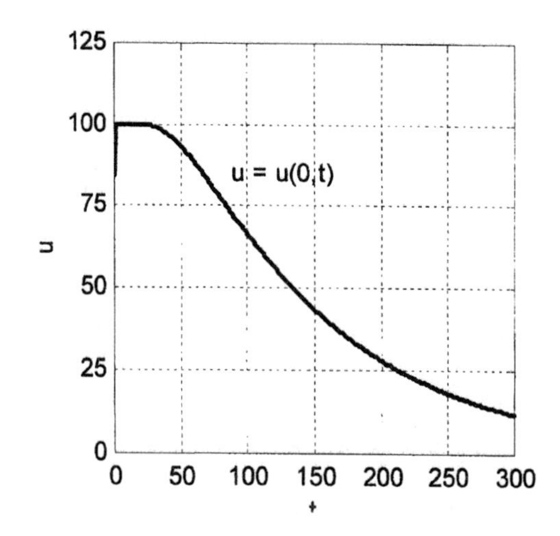

$u = u(0,t)$ for a 5-minute period shown in the right-hand figure indicates that the center-line temperature takes a bit more than 200 seconds to fall to 25°.

Investigation A

For your personal cylindrical rod with constant initial temperature $u(r,0) \equiv u_0$ to investigate in the manner we illustrate, let $c = 2p$ and $k = (0.1)q$, where p is the largest and q the smallest nonzero digit of your student ID number. If the cylindrical boundary of the rod is held at zero temperature $u(c,t) = 0$, plot graphs as above, and then determine how long it will take for the rod's center-line temperature to fall to 25°.

Investigation B

Now suppose that heat transfer occurs at the rod's cylindrical boundary, so that (according to Eq. (34) in Section 10.4) the coefficients in the series in (1) are given by

$$a_n = \frac{2u_0\gamma_n J_1(\gamma_n)}{(\gamma_n^2 + h^2)J_0(\gamma_n)^2}. \tag{4}$$

Assume that $h = 1$ so the $\{\gamma_n\}_1^\infty$ are now the positive roots of the equation

$$J_0(x) + x\,J_0'(x) = J_0(x) - x\,J_1(x) = 0 \tag{5}$$

(because $J_0'(x) = -J_1(x)$). The figure below shows the graph of the left-hand side in (5)

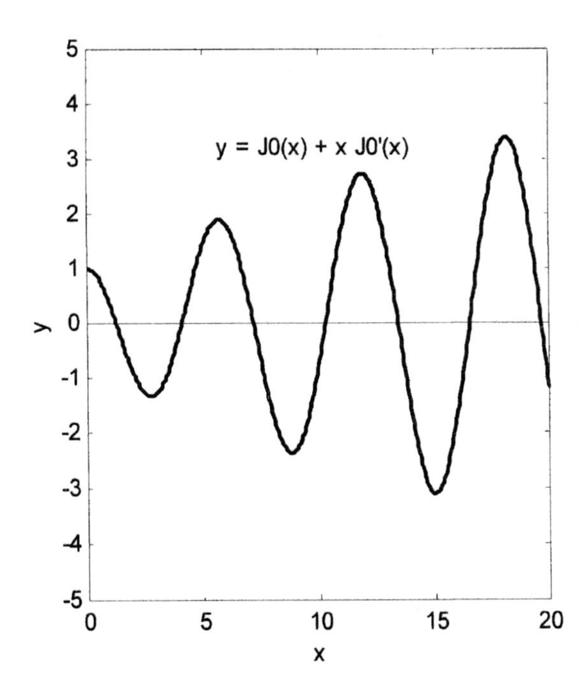

and indicates that $\gamma_1 \approx 1.25$, with successive roots differing (as usual) by approximately π. Determine in this case *both* how long it takes the center-line temperature to fall to 25° *and* how long it takes the boundary temperature at $r = c$ to fall to 25°.

Using *Maple*

Knowing that $\gamma_1 \approx 2.4$ and that successive roots of the eigenfunction equation

```
eq :=   BesselJ(0,x) = 0:
```

differ approximately by an integral multiple of π, we can approximate the first $N = 20$ values of γ_n by using the commands

```
N := 20:
g := array(1..N):    # g for gamma
for n from 1 to N do
    g[n] := fsolve(eq, x=2.4+(n-1)*Pi):
    od:
seq(g[n], n=1..N);
```

2.404825558, 5.520078110, 8.653727913, 11.79153444, 14.93091771, 18.07106397, 21.21163663, 24.35247153, 27.49347913, 30.63460647, 33.77582021, 36.91709835, 40.05842576, 43.19979171, 46.34118837, 49.48260990, 52.62405184, 55.76551076, 58.90698393, 62.04846919

Then the first N coefficients in (1) are then given by

```
c := 10:        u0 := 100:      k := 0.15:
a := array(1..m):
for n from 1 to m do
    a[n] := 2*u0/(g[n]*BesselJ(1,g[n])):
    od:
```

and the following *Maple* function sums the corresponding terms of the series.

```
n := 'n':
u := (r,t) -> sum(a[n]*exp(-g[n]^2*k*t/c^2)*
                  BesselJ(0,g[n]*r/c), n=1..N):
```

The commands

```
plot(u(r,120), r=0..c);
```

and

```
plot(u(0,t), t=0..300);
```

then construct the graphs $u = u(r,120)$ and $u = u(0,t)$ shown previously. Finally, the computation

```
fsolve(u(0,t) = 25, t = 200..250);
```

214.1025

shows that it takes about 214 seconds for the cylinder's center-line temperature to fall to 25°.

Using *Mathematica*

Knowing that $\gamma_1 \approx 2.4$ and that successive roots of the eigenfunction equation

```
eq =    BesselJ[0,x] == 0;
```

differ approximately by an integral multiple of π, we can approximate the first $M = 20$ values of γ_n by using the commands

```
M = 20;
solutions =
Table[FindRoot[eq, {x, 2.4+(n-1)*Pi}], {n,1,M}];
g = x /. solutions        (* g for gamma *)

{2.40483, 5.52008, 8.65373, 11.7915, 14.9309,
 18.0711, 21.2116, 24.3525, 27.4935, 30.6346,
 33.7758, 36.9171, 40.0584, 43.1998, 46.3412,
 49.4826, 52.6241, 55.7655, 58.9070, 62.0485}
```

Then the first m coefficients in (1) are given by

```
c = 10;     u0 = 100;     k = 0.15;
a = 2*u0/(g*BesselJ[1,g]);
```

and the following *Mathematica* function sums the corresponding terms of the series.

```
u[r_,t_] :=
Apply[Plus,
      a*Exp[-g^2*k*t/c^2]*BesselJ[0,g*r/c]] // N
```

The commands

```
Plot[Evaluate[u[r,120]], {r,0,c}];
```

and

```
Plot[Evaluate[u[0,t]], {t,0,300}];
```

then construct the graphs $u = u(r, 120)$ and $u = u(0, t)$ shown previously. Finally, the computation

```
FindRoot[u[0,t] == 25, {t,200}]
{t -> 214.013}
```

shows that it takes about 214 seconds for the cylinder's center-line temperature to fall to 25°.

Using MATLAB

First we define the eigenvalue equation $J_0(x) = 0$ in form of the function

```
j0 = inline('besselj(0,x)','x')
j0 =
     Inline function:
     j0(x) = besselj(0,x)
```

whose positive zeros $\{\gamma_n\}_1^\infty$ we need to find. Then knowing that $\gamma_1 \approx 2.4$ and that successive roots differ approximately by an integral multiple of π, we can approximate the first $m = 20$ values of γ_n by using the MATLAB commands

```
m = 20;
for  n = 1 : m
     g(n) = fzero(j0, 2.4 + (n-1)*pi );
     end

reshape(g,5,4)'
ans =
    2.4048     5.5201     8.6537    11.7915    14.9309
   18.0711    21.2116    24.3525    27.4935    30.6346
   33.7758    36.9171    40.0584    43.1998    46.3412
   49.4826    52.6241    55.7655    58.9070    62.0485
```

Now that these γ_n values are available, we can now edit the function **u.m** of Project 10.2 so as to calculate the $p \times q$ matrix **u(r,t)** of temperatures corresponding to a q-vector **r** of radii and a p-vector **t** of times:

```
function   u = u(r,t)
%  r = row vector of  q  radii of cylinder
%  t = row vector of  p  times
%  u = p x q  matrix of temperatures

c = 10;                        % radius of cylinder
```

```
u0 = 100;                    % initial temp of cylinder
k = 0.15;                    % thermal diffusivity
g = ...
  [ 2.4048     5.5201     8.6537    11.7915    14.9309 ...
   18.0711    21.2116    24.3525    27.4935    30.6346 ...
   33.7758    36.9171    40.0584    43.1998    46.3412 ...
   49.4826    52.6241    55.7655    58.9070    62.0485 ];

a = 2*u0./(g.*besselj(1,g));          %  coeffs in (1)
coeffs = diag(a);                     %  m x m  diag matrix
                                      %         of coeffs
exps = exp(-(g'.*g')*k*t/c^2);        %  m x p  matrix
bessels = besselj(0,g'*r/c);          %  m x q  matrix
u = exps'*coeffs*bessels;             %  p x q  matrix
```

The commands

```
r = 0 : 0.05 : 10;
plot(r, u(r,120))
```

and

```
t = 0 : 300;
plot(t, u(0,t))
```

then construct the graphs $u = u(r,120)$ and $u = u(0,t)$ shown previously. Finally, if we define the adjusted center-line temperature function

```
function  y = u0(t)
y = u(0,t)  - 25;
```

then the computation

```
fzero('u0',200)
ans =
   214.1071
```

shows that it takes about 214 seconds for the cylinder's center-line temperature to fall to 25°.

Application 10.5A
Heat Flow in Rectangular Plates

Example 1 in Section 10.5 of the text involves heat flow in a thin rectangular plate that occupies the plane region $0 \le x \le a$, $0 \le y \le b$. It has insulated faces and its four edges are held at temperature zero. If it has the given initial temperature function $f(x,y)$, then its temperature function $u(x,y,t)$ satisfies the boundary value problem consisting of the 2-dimensional heat equation

$$\frac{\partial u}{\partial t} = k\,\nabla^2 u = k\left(\frac{\partial^2 u}{\partial x^2} + \frac{\partial^2 u}{\partial y^2}\right) \tag{1}$$

the boundary conditions

$$u(0,y,t) = u(a,y,t) = u(x,0,t) = u(x,b,t) = 0, \tag{2}$$

and the initial temperature condition

$$u(x,y,0) \;=\; f(x,y). \tag{3}$$

If $f(x,y) \equiv u_0$, a constant, then Problem 1 in the text yields the double Fourier series solution

$$u(x,y,t) \;=\; \frac{16u_0}{\pi^2} \sum_{m\text{ odd}} \sum_{n\text{ odd}} \frac{\exp(-\gamma_{mn}^2 kt)}{mn} \sin\frac{m\pi x}{a} \sin\frac{n\pi y}{b} \tag{4}$$

where

$$\gamma_{mn}^2 \;=\; \left(\frac{m^2}{a^2} + \frac{n^2}{b^2}\right)\pi^2. \tag{5}$$

In the *Maple*-, *Mathematica*-, and MATLAB-based investigations below, we will take $a \;=\; b \;=\; \pi$, $u_0 \;=\; 100°$, and $k \;=\; 0.05$ for illustrative purposes.

Investigation

Suppose the three edges $x = 0$, $y = 0$, and $y = b$ of the plate of Example 1 in Section 10.5 are held at temperature zero (as above), but the fourth edge $x = a$ is insulated, so the boundary conditions are

$$u(0,y,t) = u(x,0,t) = u(x,b,t) = u_x(a,y,t) = 0. \tag{6}$$

If the plate's initial temperature is $u(x,y,0) \;=\; f(x,y)$, show that its temperature function is given by

$$u(x,y,t)=\sum_{m=1}^{\infty}\sum_{n=1}^{\infty} c_{mn}\exp(-\gamma_{mn}^2 kt)\sin\frac{(2m-1)\pi x}{2a}\sin\frac{n\pi y}{b} \tag{7}$$

where

$$\left(\frac{\gamma_{mn}}{\pi}\right)^2=\left(\frac{2m-1}{a}\right)^2+\left(\frac{n}{b}\right)^2 \tag{8}$$

and

$$c_{mn}=\frac{4}{ab}\int_0^a\int_0^b f(x,y)\sin\frac{(2m-1)\pi x}{2a}\sin\frac{n\pi y}{b}\,dy\,dx\,. \tag{9}$$

The result of Problem 21 in Section 9.3 of the text may be useful. If $f(x,y)\equiv u_0$ (constant), show that (7) yields

$$u(x,y,t)=\frac{16u_0}{\pi^2}\sum_{m=1}^{\infty}\sum_{n\text{ odd}}\frac{\exp(-\gamma_{mn}^2 kt)}{(2m-1)n}\sin\frac{(2m-1)\pi x}{2a}\sin\frac{n\pi y}{b}. \tag{10}$$

For your personal plate to investigate numerically, take $u_0 = 100$, $a = 10p$, $b = 10q$, $k = r/10$ where p and q are the two largest digits and r is the smallest nonzero digit in you student ID number. Plot $u = u(x,y,t)$ as a function of x and y for typical values of t to verify that each such graph is symmetric with respect to the midline $y = b/2$ of the plate, so it follows (why?) that the maximum temperature in the plate at time t occurs at some point of this midline. Then determine

- How long it takes for the maximum temperature on the edge $x = a$ to fall to 20°.
- What is then the maximum interior temperature in the plate.

Using *Maple*

We define the temperature function $u(x,y,t)$ in (4) starting with the diffusivity and initial temperature

```
k :=0.5:     u0 := 100:
```

The dependence of the exponential factor in (4) on the indices m and n is given by

```
c := (m,n) -> sqrt(m^2 + n^2):
exps := (m,n,t) -> exp(-c(m,n)^2*k*t)/(m*n):
```

Then our partial sum using the first

```
N := 8:
```

odd indices in each dimension is defined by

```
u := (x,y,t) ->
     (16*u0/Pi^2)*sum(sum(exps(2*m-1,2*n-1,t)*
                    sin((2*m-1)*x)*sin((2*n-1)*y),
                           m=1..N), n=1..N):
```

For instance, at time $t = 10$ the temperature at the center point $(\pi/2, \pi/2)$ of our $\pi \times \pi$ square plate is given by

```
Digits := 8:
evalf(u(Pi/2,Pi/2,10));
                              58.912525
```

With $N = 8$, we have summed only the terms for $m, n = 1, 3, 5, 7, 9, 11, 13, 15$ in (4), but you can check that with $N = 16$ (for instance) we get the same numerical result accurate to the decimal places shown. The command

```
plot( u(x,Pi/2,10), x=0..Pi );
```

generates the graph $u = u(x, \pi/2, 10)$ which shows (below) the expected symmetry about the center point, where the maximum temperature apparently is attained. Finally, the commands

```
Digits := 6:
[maximize(u(x,Pi/2,10),{x},{x=1..2}),
 maximize(u(Pi/2,y,10),{y},{y=1..2})];
                         [58.9125,58.9125]
```

corroborate the fact that the maximum value occurs at the center point of the plate. You probably don't want to maximize u simultaneously as a function of two variables,

```
maximize(u(x,y,10),{x,y},{x=1..2,y=1..2});
```

unless you have a very fast computer.

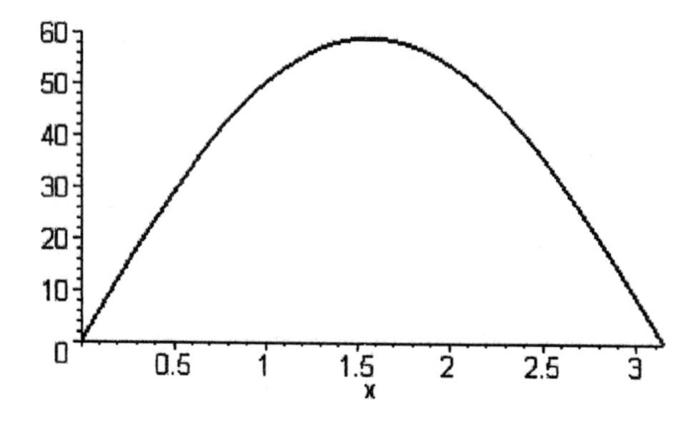

Using *Mathematica*

We define the temperature function $u(x, y, t)$ in (4) starting with the diffusivity and initial temperature

```
k = 0.05;        u0 = 100;
```

The dependence of the exponential factor in (4) on the indices m and n is given by

```
exps[m_,n_,t_] := Exp[-(m^2+n^2) k t]/(m*n)
```

Then our partial sum using maximal *odd* index

```
M = 15;
```

in each dimension is given by

```
u[x_,y_,t_] :=
 (16*u0/Pi^2) Sum[Sum[exps[m,n,t] Sin[m x] Sin[n y],
                 {m,1,M,2}], {n,1,M,2}]
```

For instance, at time $t = 10$ the temperature at the center point $(\pi/2, \pi/2)$ of our $\pi \times \pi$ square plate is given by

```
u[Pi/2, Pi/2, 10] // N

58.9125
```

With $N = 15$, we have summed only the terms for $m, n = 1, 3, 5, 7, 9, 11, 13, 15$ in (4), but you can check that with $N = 25$ (for instance) we get the same numerical result accurate to 6 significant digits as shown. The command

```
Plot3D[ Evaluate[u[x,y,10]], {x,0,Pi}, {y,0,Pi},
        PlotPoints -> {21,21},
        Shading -> False,
        BoxRatios -> {1,1,.65},
        AxesLabel -> {"x","y","u"} ];
```

generates the graph $u = u(x, y, 10)$, which shows (at the top of the next page) the expected symmetry about the center point of the plate, where the maximum temperature at time $t = 10$ evidently is attained. Finally, the calculation

```
FindMinimum[-u[x,y,10],{x,1,2},{y,1,2}]

{-58.9125, {x -> 1.5708, y -> 1.5708}}
```

corroborates the fact that the plate's maximum temperature occurs at its center point.

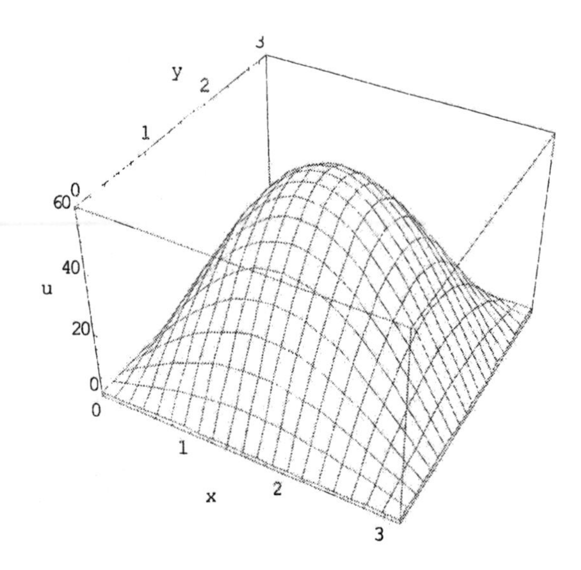

Using MATLAB

The following function **u(x,y,t)** exploits the high-dimensional array capabilities of MATLAB to accept a p-vector **x**, a q-vector **y**, and an r-vector **t** and return a 3-dimensional $p \times q \times r$ array **u** of temperatures.

```
function  u = u(x,y,t)
u0 = 100;                  % initial temperature
k = 0.05;                  % thermal diffusivity
N = 15;                    % maximal odd index
m = 1 : 2 : N;             % odd indices
n = m;
[X,Y,T,M,N] = ndgrid(x,y,t,m,n);
coeffs = (16*u0/pi^2) ./(M.*N);
exps = exp(-(M.^2+N.^2)*k.*T);
sin_mx = sin(M.*X);
sin_ny = sin(N.*Y);
terms = coeffs.*exps.*sin_mx.*sin_ny;
u = sum(sum(terms,5),4);
```

Internally, **X,Y,T,M,N** are 5-dimensional arrays with **X** consisting of copies of **x** arranged along its 1st dimension, **Y** consisting of copies of **y** arranged along its 2nd dimension, and so on. The element-wise product **terms** is a 5-dimensional array with the indices m and n varying along its 4th and 5th dimensions. The individual terms are summed with respect to the last two indices in the last line.

Of course, we can enter scalars (length 1 vectors) if we wish. For instance, at time $t = 10$ the temperature at the center point $(\pi/2, \pi/2)$ of our $\pi \times \pi$ square plate is given by

```
u(pi/2,pi/2,10)
ans =
   58.9125
```

We can plot u as a function of x and z after 10 seconds with the commands

```
x = 0 : pi/20 : pi;
y = x;
mesh(x,y,u(x,y,10))
```

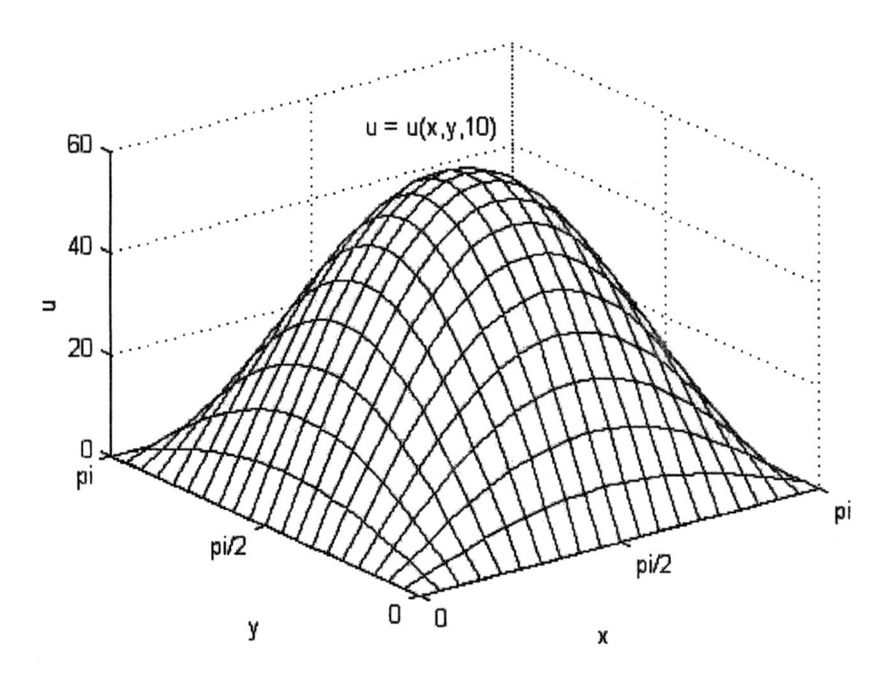

We can plot the plate's center-point temperature $u(\pi/2, \pi/2, t)$ for $0 \le t \le 50$ with the commands

```
t = 0 : 0.2 : 50;
plot(t, squeeze(u(pi/2,pi/2,t)))
```

where **squeeze** is used to collapse the $1 \times 1 \times 251$ array of temperatures to a simple vector of temperatures. Looking at the resulting graph (on the next page), we see that the plate's center-point temperature falls to 20° in a bit over 20 seconds. In order to use **fzero** to approximate this time accurately, we express the difference between 20 and u at $(\pi/2, \pi/2)$ as a function of the single variable t:

```
function  v = uc(t)
v = u(pi/2,pi/2,t) - 20;
```

Then the computation

```
fzero('uc',20)
ans =
   20.9241
```

shows that the center-point temperature hits 20° after about 20.92 seconds.

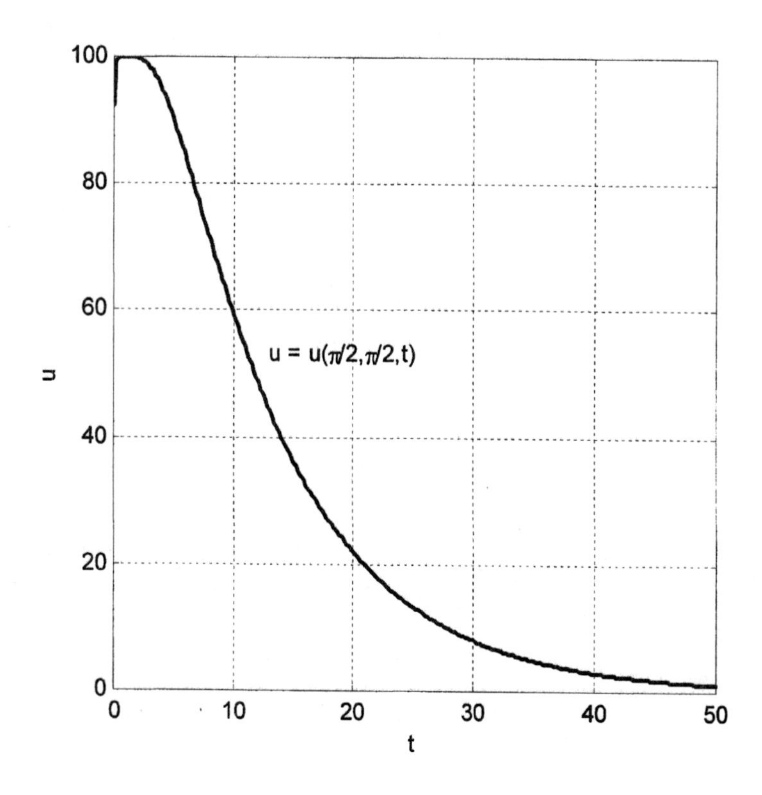

Further Investigation

As with Application 9.5, we can use *Mathematica* in other ways to gain further insight into the temperature function $u(x,y,t)$; Maple and MATLAB users can substitute equivalent commands. After the *Mathematica* commands given above to define $u(x,y,t)$ have been executed, the following command will create an animation showing the graph of $u(x,y,t)$ for values of t ranging from 1 to 20 in increments of 0.25. Double-click on any of the frames to play the animation; during play you can use the number keys 1,2,3,... to adjust the speed of play:

```
Table[Plot3D[Evaluate[u[x, y, t]],
        {x, 0, Pi}, {y, 0, Pi},
        PlotRange -> {0, 100}, Shading -> False,
        BoxRatios -> {1, 1, 0.65},
        AxesLabel -> {"x", "y", "u"},
        ImageSize -> 6*72], {t, 1, 20, 0.25}];
```

Play the animation slowly and answer these questions:

- What parts of the plate seem initially to cool off the most rapidly? The most slowly? Why might this be?

- At time $t = 1$ (the first frame of the animation), would you say that the temperature near the center of the plate is fairly uniform, or is the center point clearly hotter than the surrounding area? It might be helpful to plot the cross section of $u(x, y, 1)$ corresponding to $y = \pi/2$:

```
Plot[u[x, Pi/2, 1], {x, 0, Pi}, PlotRange -> All]
```

- Now consider the same question regarding the middle of the animation—does your answer change?

We can also illustrate—and gain insight into—the cooling of this plate through the use of color. The following **Do** loop will generate an animation that portrays the "hot" areas of the rod in red and the "cool" areas in blue for the same values of *t* as above:

```
Do[
     ContourPlot[Evaluate[u[x, y, t]],
     {x, 0, Pi}, {y, 0, Pi},
     AspectRatio -> Automatic, ImageSize -> 6*72,
     ContourLines -> False, Contours -> 150,
     ColorFunction -> (Hue[0.7(1 - #/100)]&),
     FrameTicks -> {Automatic, None},
     ColorFunctionScaling -> False],
{t, 1, 20, 0.25}];
```

Roughly speaking, what shape does the initial hot area seem to have initially? Yet as the plate cools, what shape does the warm area seem to take on? Is this a coincidence?

Application 10.5B
Rectangular Membrane Vibrations

Here we investigate the vibrations of a flexible membrane whose equilibrium position is the rectangle $0 \le x \le a,\ 0 \le y \le b$. Suppose it is released from rest with given initial displacement, and thereafter its four edges are held fixed. Then (under the usual assumptions) its displacement function $u(x,y,t)$ satisfies the boundary value problem

$$\frac{\partial^2 u}{\partial t^2} = c^2\left(\frac{\partial^2 u}{\partial x^2}+\frac{\partial^2 u}{\partial y^2}\right) \qquad \left(c^2 = T/\rho\right) \tag{1}$$

$$u(0,y,t) = u(a,y,t) = u(x,0,t) = u(x,b,t) = 0 \tag{2}$$

$$u(x,y,0) = f(x,y), \quad u_t(x,y,0) = 0. \tag{3}$$

According to Problem 3 in Section 10.5 of the text, the solution is given by

$$u(x,y,t) = \sum_{m=1}^{\infty}\sum_{n=1}^{\infty} c_{mn}\sin\frac{m\pi x}{a}\sin\frac{n\pi y}{b}\cos\gamma_{mn}ct \tag{4}$$

where the coefficients are defined by

$$c_{mn} = \frac{4}{ab}\int_0^a\int_0^b f(x,y)\sin\frac{m\pi x}{a}\sin\frac{n\pi y}{b}\,dy\,dx. \tag{5}$$

The *mn*th term in (4) corresponds to the membrane's *mn*th natural mode of oscillation with displacement function

$$u_{mn}(x,y,t) = \sin\frac{m\pi x}{a}\sin\frac{n\pi y}{b}\cos\gamma_{mn}ct \tag{6}$$

with circular frequency $\omega_{mn} = \gamma_{mn}c$ where

$$\gamma_{mn}^2 = \left(\frac{m^2}{a^2}+\frac{n^2}{b^2}\right)\pi^2. \tag{7}$$

The *mn*th inital position function

$$u_{mn}(x,y) = \sin\frac{m\pi x}{a}\sin\frac{n\pi y}{b} \tag{8}$$

is the rectangular membrane's ***mn*th eigenfunction.**

Investigation

For simplicity, take $a = b = c = 1$ and plot some eigenfunctions with small values of m and n in (8). Then plot linear combinations of several eigenfunctions to see some of the more interesting possible initial shapes of a vibrating membrane. For example, the figure below shows the graph of the initial position function

$$u(x,y) = u_{11}(x,y) - 3\,u_{22}(x,y) \;=\; \sin x \sin y - 3 \sin 2x \sin 2y$$

generated by the *Mathematica* commands

```
u = Sin[x] Sin[y]  - 3 Sin[2x] Sin[2y];
Plot3D[ Evaluate[u], {x,0,Pi}, {y,0,Pi},
        PlotPoints -> {20,20},
        Shading -> False,
        ViewPoint -> {-1.5,3,0.5} ]
```

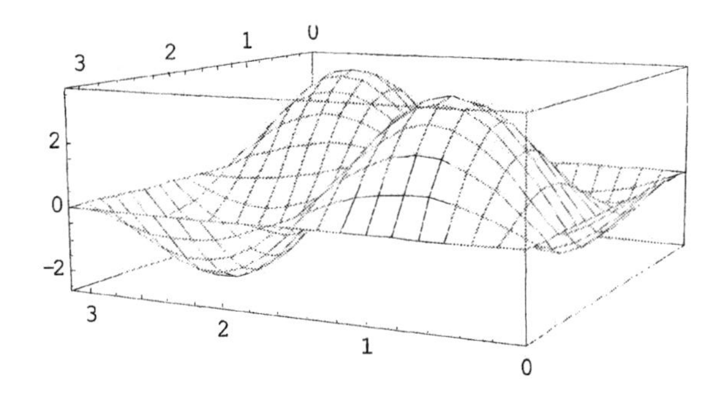

The *Maple* commands

```
u := sin(x)*sin(y)  - 3*sin(2*x)*sin(2*y):
plot3d(u, x=0..Pi, y=0..Pi);
```

and the MATLAB commands

```
x = 0 : pi/20 : pi;      y = x;
[x,y] = meshgrid(x,y);
u = sin(x).*sin(y)  - 3*sin(2*x).*sin(2*y);
surf(x,y,u)
```

produce similar results.

Maple, *Mathematica*, and MATLAB all have the capability to animate a sequence of snapshots of a vibrating membrane so as to show a "movie" illustrating its motion. For instance the *Maple* commands

```
u  := (x,y) -> sin(x)*sin(2*y) + sin(2*x)*sin(y):

w := sqrt(5):       # circular frequency
p := 2*Pi/w:        # period of oscillation

with(plots):
animate3d( u(x,y)*cos(w*t),
           x=0..Pi, y=0..Pi, t=0..p,
           frames=12, style = patch );
```

produce a 12-frame movie showing one complete oscillation of the membrane with initial position function

$$u(x, y) = \sin x \sin 2y + \sin 2x \sin y \qquad (9)$$

and circular frequency $\omega = \sqrt{5}$. The *Mathematica* commands

```
u = Sin[x] Sin[2y] + Sin[2x] Sin[y];

w = Sqrt[5];       (* circular frequency    *)
P = 2 Pi/w;        (* period of oscillation *)

frame = Table[ Plot3D[ Evaluate[u Cos[w t]],
               {x,0,Pi}, {y,0,Pi},
               PlotRange -> {-1.5, 1.5},
               BoxRatios -> {3,3,2},
               ViewPoint->{-1.5, 2.8, 0.75} ],
               {t,0,P/2, P/20 } ];
```

produce a movie of a half-oscillation which (with the **Animate Graphics** selection) can be played back-and-forth to show successive oscillations continuously. The command

```
Show[ GraphicsArray[ {{frame[[1]],  frame[[ 3]]},
                      {frame[[5]],  frame[[ 7]]},
                      {frame[[9]],  frame[[11]]}} ]]
```

displays the array of successive snapshots shown on the next page. Corresponding MATLAB commands are included in the m-file for this project that can be downloaded from the DE projects page at the web site **`www.prenhall.com/edwards`**.

Experiment in this way with linear combinations of two, three, or more membrane eigenfunctions of the form

$$u_{mn}(x, y, t) = \sin mx \sin ny \cos \omega_{mn} t \qquad (10)$$

where $\omega_{mn} = \sqrt{m^2 + n^2}$. Vary the coefficients so as to produce a visually attractive movie of sufficient complexity to be interesting.

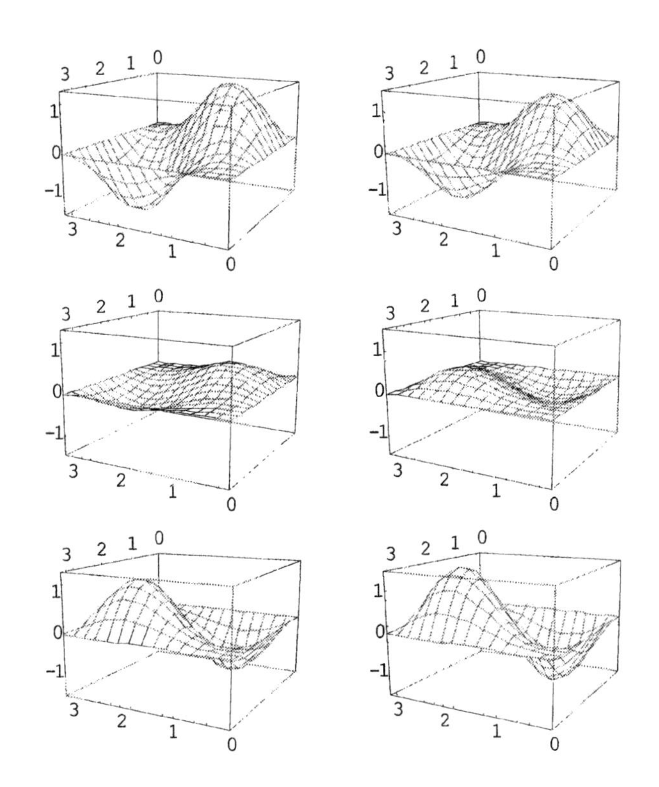

The Plucked Square Membrane

Suppose the square membrane $0 \le x, y \le \pi$ is plucked at its center point and set in motion from rest with the initial position function

$$u(x, y, 0) = f(x, y) = \min(x, y, \pi - x, \pi - y) \tag{11}$$

whose graph over the square $0 \le x, y \le \pi$ looks like a square tent or pyramid with height $\pi/2$ at its center. Thus the "tent function" $f(x, y)$ is the 2-dimensional analogue of the familiar 1-dimensional triangle function that describes the initial position of a plucked string. It can be defined "piecewise" as indicated in the figure on the next page. This diagram indicates how to subdivide the domain of definition of the function f in the integral in (5) — with $a = b = \pi$ — in order to calculate the coefficients $\{c_{mn}\}$ in (4).

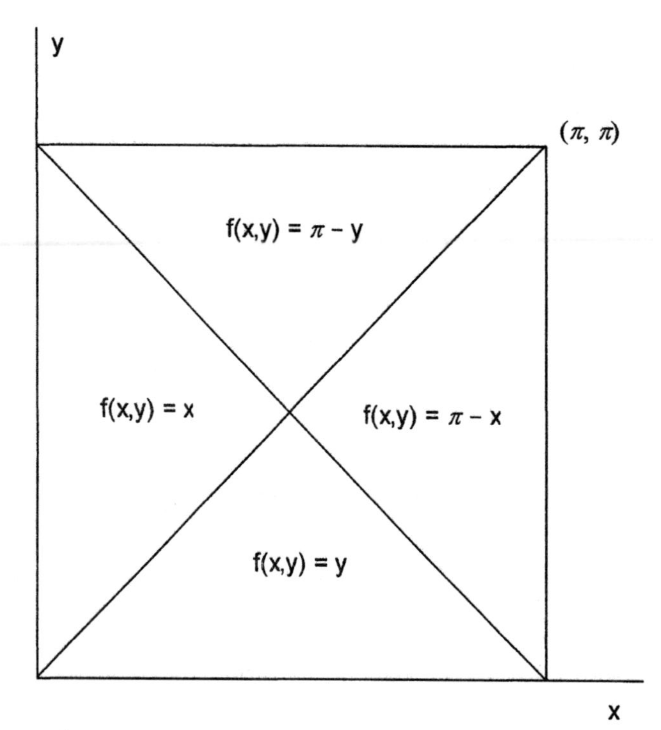

Evidently we can write

$$f(x,y) = \begin{cases} x & \text{if } x < y < \pi - x,\ 0 < x < \pi/2, \\ y & \text{if } y < x < \pi - y,\ 0 < y < \pi/2, \\ \pi - y & \text{if } \pi - y < x < y,\ \pi/2 < y < \pi, \\ \pi - x & \text{if } \pi - x < y < x,\ \pi/2 < x < \pi. \end{cases}$$

We proceed here with a *Maple*-based investigation of the motion of the square membrane if it starts from rest with the initial position function defined in (11). Analogous *Mathematica*- and MATLAB-based investigations can downloaded from the DE computing projects web site mentioned previously.

The coefficient integral in (5) is the sum of four double integrals corresponding to the four triangles in the figure above. To evaluate these integrals symbolically, we enter the *Maple* commands

```
I1 := int( int(x*sin(m*x)*sin(n*y),
              y=x..Pi-x), x=0..Pi/2):
I2 := int( int(y*sin(m*x)*sin(n*y),
              x=y..Pi-y), y=0..Pi/2):
I3 := int( int((Pi-y)*sin(m*x)*sin(n*y),
              x=Pi-y..y), y=Pi/2..Pi):
I4 := int( int((Pi-x)*sin(m*x)*sin(n*y),
              y=Pi-x..x), x=Pi/2..Pi):
```

Then the sum in (5) is given by

```
c := simplify((4/Pi^2)*(I1+I2+I3+I4));
```

$$c := 4\frac{-n\sin(\pi n) + n\cos(\pi m)\sin(\pi n) + \sin(\pi m)\,m - \sin(\pi m)\cos(\pi n)\,m}{n\,m\,\pi^2\,(-m^2+n^2)}$$

We see (from the denominator here) that *Maple* is assuming m and n not equal, in which case it is obvious that $c_{mn} = 0$ because of the sine factors.

To calculate the non-zero "diagonal coefficients" in the Fourier series (4), we repeat the computation above with $m = n$ from the beginning. The result is

$$c_{nn} = \frac{2\left[1-(-1)^n\right]}{\pi n^2} = \begin{cases} 4/\pi n^2 & \text{for } n \text{ odd,} \\ 0 & \text{for } n \text{ even.} \end{cases} \tag{12}$$

Thus the Fourier series of the tent function $f(x,y)$ defined in (11) is

$$f(x,y) = \frac{4}{\pi}\sum_{n\ \mathrm{odd}}^{\infty}\frac{\sin nx\,\sin ny}{n^2}. \tag{13}$$

It follows that the solution of our original vibrating membrane problem with initial position function $f(x, y)$ is given by

$$u(x,y,t) = \frac{4}{\pi}\sum_{n\ \mathrm{odd}}^{\infty}\frac{\sin nx\,\sin ny\,\cos nt\sqrt{2}}{n^2}. \tag{14}$$

We invite you check out the computation outlined here using either *Maple*, *Mathematica*, or MATLAB. Try unequal values of a and b to see whether you still get a "diagonal" series as in (14).

Is it now clear — because (14) contains *no* terms with $m \neq n$ — that the function $u(x,y,t)$ is **periodic** (in t) with period $P = \pi\sqrt{2}$? This fact, that the tent function in (14) thus yields a "musical" vibration of a square membrane, was first pointed out to us by John Polking of Rice University.

To investigate this vibration visually, we proceed to define a partial sum of the series in (14).

```
c := n -> 4/(Pi*n^2):          # with n odd
N := 13:                       # so 2N-1 = 25
P := Pi*sqrt(2):               # period of oscillation
u := (x,y,t) ->
     sum(c(2*k-1)*sin((2*k-1)*x)*
     sin((2*k-1)*y)*cos((2*k-1)*P*t/Pi), k=1..N):
```

The following commands plot now snapshots of the resulting vibration with $t=0$ and with $t=P$.

```
plot3d(u(x,y,0),  x=0..Pi, y=0..Pi);
plot3d(u(x,y,P/8),x=0..Pi, y=0..Pi);
```

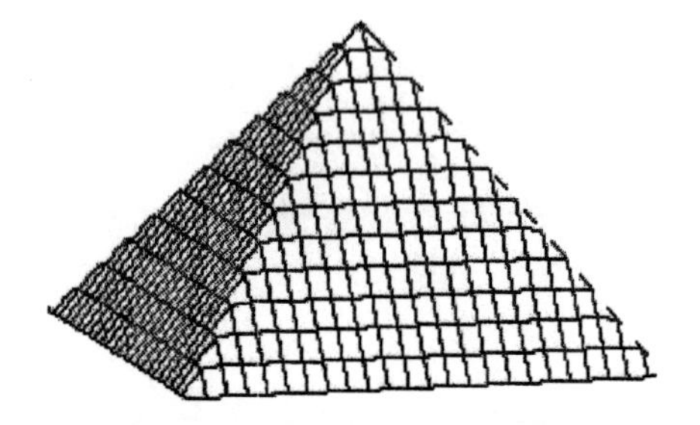

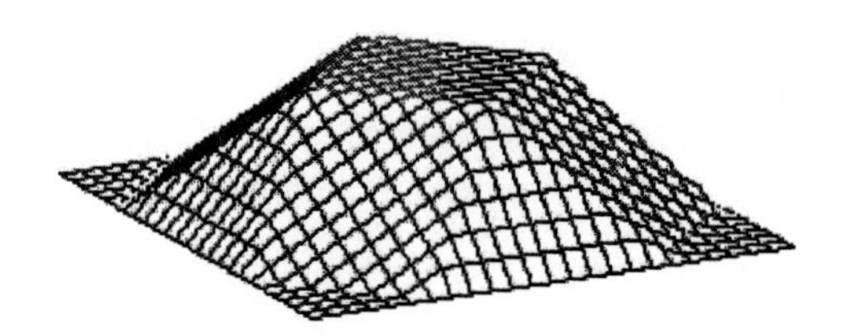

Thus a vibrating plucked membrane exhibits a "flat spot" that is reminiscent of the flat spot we see in vibrations of a plucked string. Finally, the commands

```
with(plots):
animate3d( u(x,y,t),
           x=0..Pi, y=0..Pi, t=0..P,
           frames=17, style = patch );
```

construct a17-frame movie that you can play (using the **Animate** menu) — either continuously or one-frame-at-a-time — to investigate the periodic oscillation. For instance, you will find that the 8th frame (after one-half oscillation) shows a pyramid "pointed" downward instead of upward.

Project 10.5C
Circular Membrane Vibrations

In problems involving regions that enjoy circular symmetry about the origin in the plane (or the vertical z-axis in space), the use of polar (or cylindrical) coordinates is advantageous. In Section 9.7 of the text we discussed the expression of the 2-dimensional Laplacian

$$\nabla^2 u = \frac{\partial^2 u}{\partial r^2} + \frac{1}{r}\frac{\partial u}{\partial r} + \frac{1}{r^2}\frac{\partial^2 u}{\partial \theta^2} \tag{1}$$

in terms of the familiar plane polar coordinates (r,θ) for which $x = r\cos\theta$ and $y = r\sin\theta$. If $u(r,\theta,t)$ denotes the vertical displacement at time t of the point (r,θ) of a vibrating circular membrane of radius c, then the 2-dimensional wave equation takes the polar coordinate form

$$\frac{\partial^2 u}{\partial t^2} = a^2\nabla^2 u = a^2\left(\frac{\partial^2 u}{\partial r^2} + \frac{1}{r}\frac{\partial u}{\partial r} + \frac{1}{r^2}\frac{\partial^2 u}{\partial \theta^2}\right). \tag{2}$$

where $a^2 = T/\rho$ in terms of the membrane's tension T and density ρ (per unit area). If the membrane is released from rest with given initial position function $f(r,\theta)$ at time $t = 0$ and thereafter its boundary is held fixed, then the membrane's displacement function $u(r,\theta,t)$ satisfies both (2) and the boundary conditions

$$u(c,\theta,t) = 0, \qquad \text{(fixed boundary)} \tag{3}$$

$$u(r,\theta,0) = f(r,\theta), \qquad \text{(given initial displacement)} \tag{4}$$

$$u_t(r,\theta,0) = 0. \qquad \text{(zero initial velocity)} \tag{5}$$

Fill in the details in the solution that is outlined as follows. Show first that the substitution

$$u(r,\theta,t) = R(r)\,\Theta(\theta)\,T(t) \tag{6}$$

in the wave equation (2) yields the separation of variables

$$\frac{T''}{a^2 T} = \frac{R'' + \frac{1}{r}R'}{R} + \frac{\Theta''}{r^2\Theta} = -\alpha^2 \qquad \text{(constant).} \tag{7}$$

Then

$$T'' + \alpha^2 a^2 T = 0, \qquad T'(0) = 0 \tag{8}$$

implies that (to within a constant multiple)

$$T(t) = \cos \alpha a t. \tag{9}$$

Next, the right equality in (43) yields the equation

$$\frac{r^2 R'' + rR'}{R} + \alpha^2 r^2 + \frac{\Theta''}{\Theta} = 0 \tag{10}$$

from which it follows that

$$\frac{\Theta''}{\Theta} = -\beta^2 \qquad \text{(constant).} \tag{11}$$

In order that a solution $\Theta(\theta)$ of $\Theta'' + \beta^2\Theta = 0$ have the necessary 2π-periodicity, the parameter β must be an integer, so we have the θ-solutions

$$\Theta_n(\theta) = \begin{cases} \cos n\theta \\ \sin n\theta \end{cases} \tag{12}$$

for $n = 0, 1, 2, 3, \ldots\ldots$.

Substitution of $\Theta''/\Theta = -n^2$ in (10) now yields the parametric Bessel equation

$$r^2 R'' + rR' + (\alpha^2 r^2 - n^2)R = 0 \tag{13}$$

of order n, with bounded solution $R(r) = J_n(\alpha r)$. Since the zero boundary condition (3) yields $J_n(\alpha c) = 0$, case 1 in the table of Fig. 10.4.2 of the text yields the r-eigenfunctions

$$R_{mn}(r) = J_n\left(\frac{\gamma_{mn} r}{c}\right) \qquad (m = 1, 2, 3, \ldots;\ n = 0, 1, 2, \ldots) \tag{14}$$

where γ_{mn} denotes the mth positive solution of the equation $J_n(x) = 0$. Finally, substitution of $\alpha_{mn} = \gamma_{mn}/c$ in (9) yields the t-function

$$T_{mn}(t) = \cos \frac{\gamma_{mn} a t}{c}. \tag{15}$$

Combining (12), (14), and (15) we see that our boundary value problem for the circular membrane released from rest has the formal series solution

$$u(r,\theta,t) = \sum_{m=1}^{\infty} \sum_{n=0}^{\infty} J_n\left(\frac{\gamma_{mn} r}{c}\right)(a_{mn} \cos n\theta + b_{mn} \sin n\theta) \cos \frac{\gamma_{mn} a t}{c}. \tag{16}$$

Thus the vibrating circular membrane's typical natural mode of oscillation with zero initial velocity is of the form

$$u_{mn}(r,\theta,t) = J_n\left(\frac{\gamma_{mn}r}{c}\right)\cos n\theta \cos\frac{\gamma_{mn}at}{c} \tag{17}$$

or the analogous form with $\sin n\theta$ instead of $\cos n\theta$. In this mode the membrane vibrates with $m-1$ fixed *nodal circles* (in addition to its boundary circle $r = c$) with radii $r_{jn} = \gamma_{jn}c/\gamma_{mn}$ for $j = 1, 2, \ldots\ldots, m-1$. It also has $2n$ fixed *nodal radii* spaced at angles of π/n starting with $\theta = \pi/(2n)$. Figure 10.5.6 in the text shows some typical configurations of these nodal circles and radii, which divide the circle into annular sectors that move alternately up and down as the membrane vibrates.

We proceed here with a *Mathematica*-based investigation of circular membrane vibrations. Analogous *Maple*- and MATLAB-based investigations can downloaded from the DE computing projects link at the web site **www.prenhall.com/edwards**.

Figure 10.5.5 in the text shows the graphs of $y = J_n(x)$ for $n = 1, 2, 3, 4$. We see there that $n+3$ is (at least for small n) a rough but reasonable initial estimate of the first positive zero γ_{1n} of the equation $J_n(x) = 0$. This observation motivates the *Mathematica* commands

```
inits =
Table[ FindRoot[ BesselJ[n,x] == 0, {x, n+3} ],
       {n, 1,5} ];
g1 = x /. inits
```

```
{3.83171, 5.13562, 6.38016, 7.58834, 8.77148}}
```

that accurately approximate these initial zeros of $J_1(x), J_2(x), J_3(x), J_4(x)$, and $J_5(x)$.

Now recall that the gap between successive zeros of $J_n(x) = 0$ is approximately π, so it follows that $\gamma_{mn} \approx \gamma_{1n} + (m-1)\pi$. Consequently the commands

```
zeros = Table[ FindRoot[ BesselJ[n,x] == 0,
               {x, g1[[n]] + (m-1)Pi} ],
               {m,1,5}, {n,1,5} ];
g = x /. zeros;
g // TableForm
```

```
3.83171   5.13562   6.38016   7.58834   8.77148
7.01559   8.41724   9.76102   11.0647   12.3386
10.1735   11.6198   13.0152   14.3725   15.7002
13.3237   14.796    16.2235   17.616    18.9801
16.4706   17.9598   19.4094   20.8269   22.2178
```

yield a table displaying the *m*th zero γ_{mn} of $J_n(x)=0$ in the *m*th row and *n*th column. For instance,

```
g[[2,3]]
9.76102
```

so we see that the 2nd zero of $J_3(x)=0$ is $\gamma_{23}\approx 9.76102$.

Now that numerical values of the zeros of Bessel functions are available, we can employ the commands

```
x = r Cos[t];  y = r Sin[t];
m = 2;    n = 1;
ParametricPlot3D[
     {x,y,BesselJ[n, g[[m,n]] r] Cos[n t]},
     {r, 0,1}, {t, 0, 2Pi} ]
```

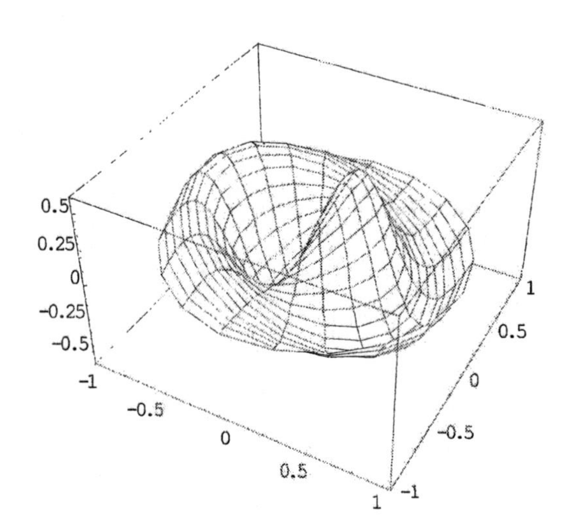

to display the initial position corresponding to an eigenfunction defined as in (17). The following commands graph these initial positions with $m = 1$ and $n = 1, 2, 3, 4$.

```
shape = Table[
         ParametricPlot3D[
            {x,y, BesselJ[n, g[[1,n]] r] Cos[n t]},
            {r, 0,1}, {t, 0, 2Pi}, Shading -> False,
            PlotPoints -> {12,48}, Ticks -> None,
            ViewPoint->{1.300, -2.400, 1.000}],
            {n,1,4} ];
```

```
Show[GraphicsArray[{{shape[[1]], shape[[2]]},
                    {shape[[3]], shape[[4]]}}]];
```

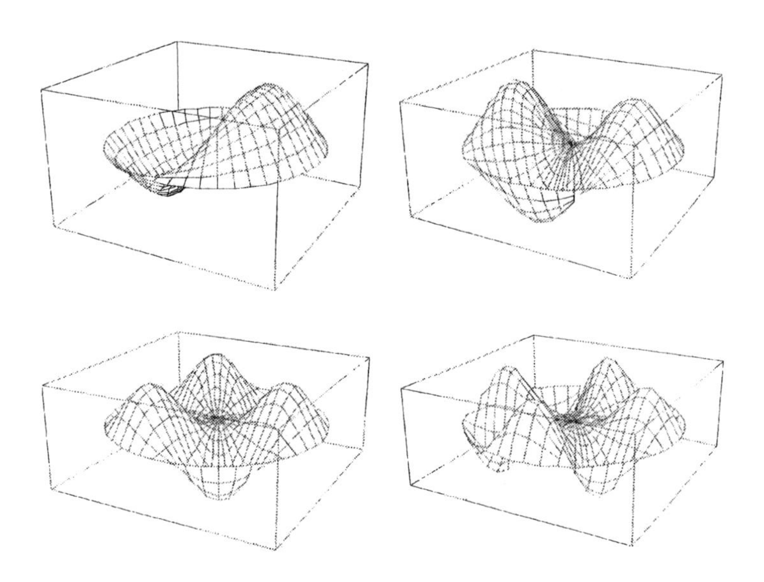

We suggest that you explore vibrating circular membrane possibilities by graphing convenient linear combinations of eigenfunctions defined as in (17). For instance, the figure below shows the initial position of the oscillation

$$u(r,\theta,t) = J_1\left(\gamma_{21}r\right)\cos\theta\cos\gamma_{21}t + J_2\left(\gamma_{32}r\right)\cos 2\theta\cos\gamma_{32}t \qquad (18)$$

defined for for a circular membrane with $a = 1$ and radius $c = 1$.

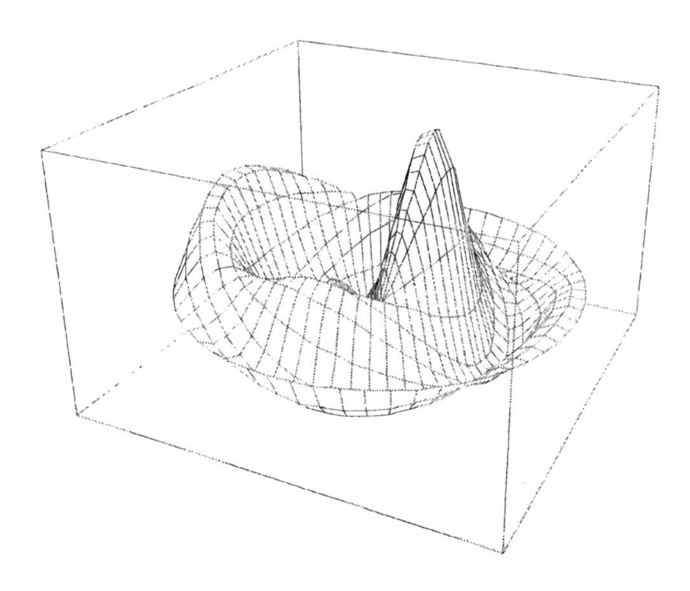

The following *Mathematica* commands generate a sequence of $k \cdot p = 80$ frames that can be animated to show a movie illustrating the circular membrane oscillation defined in (18).

```
k = 4;           (* cycles *)
p = 20;          (* frames/cycle *)

g21 = g[[2,1]];  g32 = g[[3,2]];
w1 = g21;        w2 = g32;          (* frequencies *)

x = r Cos[theta];  y = r Sin[theta];

z = BesselJ[1, g21 r] Cos[  theta] Cos[w1 t] +
    BesselJ[2, g32 r] Cos[2 theta] Cos[w2 t];

Do[ ParametricPlot3D[ Evaluate[{x,y,z}],
     {r, 0,1}, {theta, 0,2Pi},
     PlotRange -> Automatic,
     BoxRatios -> {1,1,0.6},
     ViewPoint -> {1,-2,1},
     PlotPoints -> {15,60},
     Ticks -> None,
     Shading -> False],
     {t,0, k Pi/w1, Pi/(p w1)} ];
```

Application 10.5D
Spherical Harmonic Waves

In problems involving regions that enjoy spherical symmetry about the origin in space, it is appropriate to use spherical coordinates. The 3-dimensional Laplacian for a function $u(\rho,\phi,\theta)$ expressed in spherical coordinates is given by

$$\nabla^2 u = \frac{1}{\rho^2}\left[\frac{\partial}{\partial\rho}\left(\rho^2\frac{\partial u}{\partial\rho}\right)+\frac{1}{\sin\phi}\frac{\partial}{\partial\phi}\left(\sin\phi\frac{\partial u}{\partial\phi}\right)+\frac{1}{\sin^2\phi}\frac{\partial^2 u}{\partial\theta^2}\right]. \tag{1}$$

Note that $\rho = |OP|$ denotes the distance of the point P from the origin O, ϕ is the angle down from the vertical z-axis to OP, and θ is the ordinary polar coordinate angle in the horizontal xy-plane (though some texts reverse the roles of ϕ and θ). If u is independent of either ρ, ϕ, or θ then the corresponding second-derivative term is missing on the right-hand side in (1).

For instance, consider radial vibrations of the surface of a spherical planet of radius c. If $u(\phi,\theta,t)$ denotes the radial displacement at time t of the point (ϕ,θ) of the surface $\rho = c$ of the planet, then the wave equation $u_{tt} = a^2\nabla^2 u$ takes the form

$$\frac{\partial^2 u}{\partial t^2} = b^2\,\nabla^2_{\phi\theta}u \tag{2}$$

where $b = a/c$ and

$$\nabla^2_{\phi\theta}u = \frac{1}{\sin\phi}\frac{\partial}{\partial\phi}\left(\sin\phi\frac{\partial u}{\partial\phi}\right)+\frac{1}{\sin^2\phi}\frac{\partial^2 u}{\partial\theta^2}. \tag{3}$$

Alternatively, Equation (2) models the oscillations of tidal waves on the surface of a water-covered spherical planet with radius c. In this case, $u(\phi,\theta,t)$ denotes the vertical displacement (from equilibrium) of the water surface at the point (ϕ,θ) at time t and $b^2 = gh/c^2$, where h is the average depth of the water and g denotes gravitational acceleration at the surface of the planet.

To solve equation (2) by separation of variables, show first that the substitution

$$u(\phi,\theta,t) = Y(\phi,\theta)\,T(t) \tag{4}$$

yields the equations

$$T'' + b^2\lambda T = 0, \tag{5}$$

$$\nabla^2_{\phi\theta}Y + \lambda Y = 0 \tag{6}$$

where λ is the usual separation constant. Next, show that the substitution

$$Y(\phi,\theta) = \Phi(\phi)\,\Theta(\theta) \tag{7}$$

in (6) yields the equations

$$\Theta'' + \mu\,\Theta = 0, \tag{8}$$

$$(\sin^2\phi)\,\Phi'' + (\sin\phi\cos\phi)\,\Phi' + (\lambda\sin^2\phi - \mu) = 0 \tag{9}$$

where μ is a second separation constant.

In order that $\Theta(\theta)$ be periodic with period 2π, it follows from (8) that $\mu = m^2$, the square of a non-negative integer, in which case a typical solution of (8) is

$$\Theta_m(\theta) = \cos m\theta. \tag{10}$$

Now show that with $\mu = m^2$ the substitution

$$x = \cos\phi \tag{11}$$

in (9) yields the ordinary differential equation

$$\left(1-x^2\right)\frac{d^2\Phi}{dx^2} - 2x\frac{d\Phi}{dx} + \left(\lambda - \frac{m^2}{1-x^2}\right)\Phi = 0. \tag{12}$$

Note that if $m = 0$ then (12) is a Legendre equation with dependent variable Φ and independent variable x. According to Section 8.2 in the text, this equation has a solution $\Phi(x)$ that is continuous for $-1 \le x \le 1$ provided that the parameter $\lambda = n(n+1)$ where n is a non-negative integer. In this case the continuous solution $\Phi(x)$ is a constant multiple of the *n*th degree Legendre polynomial $P_n(x)$.

Equation (12) is an **associated Legendre equation**, and it likewise has a solution $\Phi(x)$ that is continuous for $-1 \le x \le 1$ provided that the parameter $\lambda = n(n+1)$ with n being a non-negative integer. In this case the continuous solution $\Phi(x)$ is a constant multiple of the **associated Legendre polynomial**

$$P_n^m(x) = (-1)^m\,(1-x^2)^{m/2}\,P_n^{(m)}(x), \tag{13}$$

where the *m*th derivative of the ordinary Legendre polynomial $P_n(x)$ appears on the right. For instance, writing **Pmn** for $P_n^m(x)$, *Mathematica* gives

```
P00 = LegendreP[0,x]
```

1

```
P01 = LegendreP[1,x]
P11 = Sqrt[1-x^2] D[LegendreP[1,x],x]
```

x

$\sqrt{1-x^2}$

```
P02 = LegendreP[2,x]
P12 = -Sqrt[1-x^2] D[LegendreP[2,x],x]
P22 = (1-x^2) D[LegendreP[2,x],{x,2}]
```

$\frac{3x^2}{2}-\frac{1}{2}$

$-3x\sqrt{1-x^2}$

$3(1-x^2)$

Actually, the associated Legendre functions are built into *Mathematica*, with **LegendreP[n,m,x]** denoting $P_n^m(x)$.

Using *Maple*, you must first load the orthogonal polynomials package. Only the ordinary Legendre functions are immediately available, so you must implement the definition in (13).

```
with(orthopoly):
p4 := P(4,x);
```

$$p4 := \frac{35}{8}x^4 - \frac{15}{4}x^2 + \frac{3}{8}$$

```
p42 := expand((1-x^2)*diff(p4,x$2));
```

$$p24 := 60\,x^2 - \frac{15}{2} - \frac{105}{2}x^4$$

If $\mathbf{x}$ is a row vector with k elements then the MATLAB command **legendre(n,x)** yields an $(n+1)\times k$ matrix whose mth row contains the values of P_n^{m-1} at the elements of **x**. Thus, the computation

```
legendre(3, 0:1/3:1)
ans =
          0    -0.4074    -0.2593     1.0000
     1.5000     0.6285    -1.3665          0
          0     4.4444     5.5556          0
   -15.0000   -12.5708    -6.2113          0
```

shows that $P_3^2(1/3) \approx 4.4444$ and $P_3^2(2/3) \approx 5.5556$. (It would be instructive for you to deduce from (13) the exact values $P_3^2(1/3) \approx \frac{40}{9}$ and $P_3^2(2/3) \approx \frac{50}{9}$.

At any rate, given non-negative integers m and n with $m \le n$, substitution of $x = \cos\phi$ in the continuous solution (13) of Eq. (12) with $\lambda = n(n+1)$ yields the solution

$$\Phi_{mn}(\phi) = P_n^m(\cos\phi) = (\sin\phi)^m P_n^{(m)}(\cos\phi) \tag{14}$$

of Eq. (9) with $\mu = m^2$ and $\lambda = n(n+1)$. Substitution of $\lambda = n(n+1)$ in (5) yields the typical solution

$$T_n(t) = \cos\omega_n t \tag{15}$$

with frequency

$$\omega_n = b\sqrt{n(n+1)}. \tag{16}$$

Putting it all together, we get finally the eigenfunction

$$u_{mn}(\phi,\theta,t) = P_n^m(\cos\phi)\cos m\theta \cos\omega_n t \tag{17}$$

($0 \le m \le n = 1, 2, 3, \ldots$) of the wave equation in (2). The remaining eigenfunctions are obtained by (independently) replacing $\cos m\theta$ with $\sin m\theta$ and $\cos\omega_n t$ with $\sin\omega_n t$.

Big Waves on a Small Planet

With all this preparation, your task is to investigate graphically the way in which water waves slosh about on the surface of a small planet. Let us take a sphere of radius $c = 5$ with $b = 1$ in (2), and (somewhat unrealistically) consider waves of amplitude $h = 2$. The following MATLAB function **spharm(n,m)** constructs a $\phi\theta$-grid on the surface of the sphere and calculates the corresponding matrix **Y** of values of the **surface spherical harmonic** function $Y_{mn}(\phi,\theta) = P_n^m(\cos\phi)\cos m\theta$.

```
function    [Y,phi,theta] = spharm(n,m)
phi =    0 : pi/40 : pi;               % co-latitude
theta = 0 : pi/20 : 2*pi;              % polar angle
[theta,phi] = meshgrid(theta,phi);
Theta = cos(m*theta);
Phi = legendre(n, cos(phi(:,1)));
Phi = Phi(m + 1,:)';
pp = Phi;

for k = 2 : size(Phi,1)
     Phi = [Phi pp];
end;
```

```
Y = Phi.*Theta;
m = max(max(abs(Y)));
Y = Y/m;
```

The function **spshape(n,m)** then displays the corresponding initial graph $\rho = c + h\,Y_{mn}(\phi,\theta)$. The next figure shows such plots for $n = 4, 5, 6, 7$ and $m = 0, 1, 2, 3$ (respectively) generated by commands such as **spshape(6,2)**, which generates the figure which appears (in living color) on the cover of the text.

```
function    spshape(n,m)
c = 5;     h = 2;
[Y,phi,theta] = spharm(n,m);
rho = c + h*Y;
r = rho.*sin(phi);
x = r.*cos(theta);
y = r.*sin(theta);
z = rho.*cos(phi);
mesh(x,y,z)
axis([-6 6 -6 6 -6 6])
axis square, axis off
view(40,30)
colormap([0 0 0])
```

m = 0, n = 4

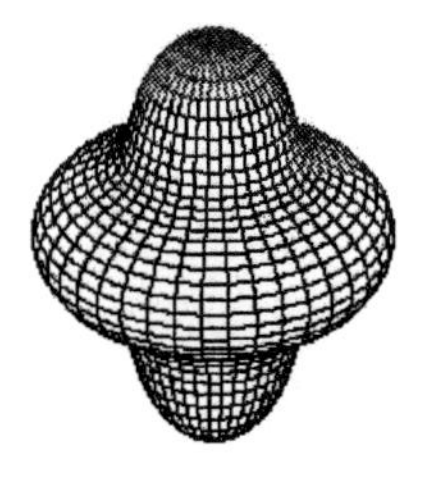

m = 1, n = 5

m = 2, n = 6

m = 3, n = 7

The following function **spmovie(n,m)** constructs and shows a movie that displays in motion the surface water waves corresponding to one of our spherical surface eigenfunctions. For instance, the command **spmovie(6,2)** sets in motion the figure on the surface of the text. *Warning*: With $k = 20$ frames, the matrix **Mslosh** storing the movie may occupies from 5 to 15 megabytes of RAM (depending upon the size of your figure window).

```
function  Mslosh = spmovie(n,m)
c = 5;
h = 2;
w = sqrt(n*(n+1));
k = 20;                          % steps per cycle
dt = 2*pi/(k*w);                 % (time) step size
[Y,phi,theta] = spharm(n,m);
% Construct the movie:
Mslosh = moviein(k);
for j = 0 : k-1
    t = j*dt;
    rho = c + h*Y*cos(w*t);
    r = rho.*sin(phi);
    x = r.*cos(theta);
    y = r.*sin(theta);
    z = rho.*cos(phi);
    surf(x,y,z)
    light
    lighting phong
    axis square, axis off
    axis([-6 6 -6 6 -6 6])
    view(40,30)
    colormap(jet)
    Mslosh(:,j+1) = getframe;
    end
% Show it:
movie(Mslosh,5)
```

Construct some movies of your own. If you're ambitious you can investigate linear combinations of different spherical surface harmonics. For instance, the spherical wave motion with initial position function

$$u(\phi,\theta,0) = 5 + 2\,P_4^1(\cos\phi)\cos\theta + P_6^3(\cos\phi)\cos 3\theta\,.$$

may remind you (at least vaguely) of a throbbing, beating human heart.

Maple- and *Mathematica*-based investigations corresponding to the MATLAB exposition here can downloaded from the DE computing projects web page at the site **www.prenhall.com/edwards**.